Department of Defense Training for
Operations with Interagency, Multinational, and Coalition Partners

Michael Spirtas · Jennifer D. P. Moroney · Harry J. Thie
Joe Hogler · Thomas-Durell Young

Prepared for the Office of the Secretary of Defense
Approved for public release; distribution unlimited

 NATIONAL DEFENSE RESEARCH INSTITUTE

The research described in this report was prepared for the Office of the Secretary of Defense (OSD). The research was conducted in the RAND National Defense Research Institute, a federally funded research and development center sponsored by the OSD, the Joint Staff, the Unified Combatant Commands, the Department of the Navy, the Marine Corps, the defense agencies, and the defense Intelligence Community under Contract W74V8H-06-C-0002.

Library of Congress Cataloging-in-Publication Data is available for this publication.

ISBN 978-0-8330-4504-1

Cover photo credits (clockwise from top left):
Army Staff Sgt. John Thomas, center, is surrounded by Afghan National Army trainers as he evaluates a target in Kandahar, Afghanistan, on January 23, 2008. (Courtesy of defenseimagery.mil, David M. Votroubek, photographer).

U.S. Navy sailors and Cameroon Naval Forces sailors unload medical supplies donated by Project Handclasp from amphibious dock landing ship USS Fort McHenry (LSD 43) to a landing craft for transport to Limbe Naval Base, Cameroon, on February 28, 2008. (Courtesy of defenseimagery.mil, Bryan A. Goyak, photographer).

Afghan men at a meeting with U.S. Army civil affairs team, on March 27, 2003. (AP Photo/Gurinder Osan).

A U.S. soldier monitors battlefield conditions at a joint U.S./Afghan military command center on June 21, 2007. (AP Photo/Musadeq Sadeq)

The RAND Corporation is a nonprofit research organization providing objective analysis and effective solutions that address the challenges facing the public and private sectors around the world. RAND's publications do not necessarily reflect the opinions of its research clients and sponsors.

RAND® is a registered trademark.

Published 2008 by the RAND Corporation
1776 Main Street, P.O. Box 2138, Santa Monica, CA 90407-2138
1200 South Hayes Street, Arlington, VA 22202-5050
4570 Fifth Avenue, Suite 600, Pittsburgh, PA 15213-2665
RAND URL: http://www.rand.org/
To order RAND documents or to obtain additional information, contact
Distribution Services: Telephone: (310) 451-7002;
Fax: (310) 451-6915; Email: order@rand.org

Preface

Current challenges that the United States is facing require U.S. forces to work with a wide range of organizations outside of the U.S. Department of Defense (DoD). This book documents research on integrated operations to provide U.S. defense planners with recommendations to better prepare U.S. military personnel to work successfully with a host of partners, such as other U.S. government (USG) agencies, multinational organizations, and coalition partner countries. This book will assist the Office of the Secretary of Defense (OSD) in thinking through its options to make effective use of limited resources available for integrated-operations training in an environment of limited resources.

This research was sponsored by the Office of the Under Secretary of Defense for Personnel and Readiness (P&R) and was conducted within the International Security and Defense Policy Center of the RAND National Defense Research Institute, a federally funded research and development center sponsored by the Office of the Secretary of Defense, the Joint Staff, the Unified Combatant Commands, the Department of the Navy, the Marine Corps, the defense agencies, and the defense Intelligence Community.

For more information on RAND's International Security and Defense Policy Center, contact the Director, James Dobbins. He can be reached by email at James_Dobbins@rand.org; by phone at 703-413-1100, extension 5134; or by mail at RAND, 1200 South Hayes Street, Arlington, VA 22202-5050. More information about RAND is available at www.rand.org.

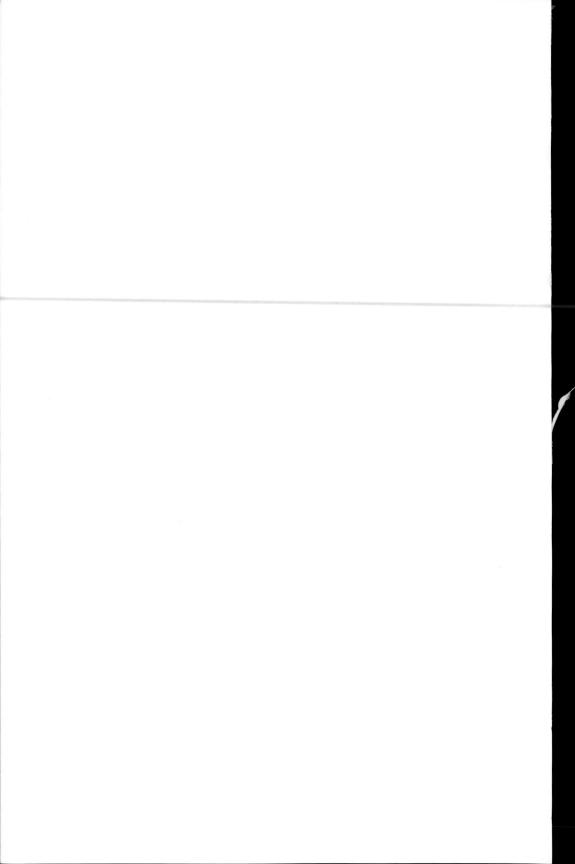

Contents

Figures

Tables

Summary

This book provides suggestions for how the U.S. military can help prepare its personnel to work successfully with interagency (IA), multinational, and coalition partners. The nature of recent challenges and the types of missions the U.S. Department of Defense (DoD) has undertaken highlight the need for DoD to consider ways to help the military prepare to work with other government agencies, international organizations, private and nongovernmental organizations, and foreign militaries. These challenges require DoD to combine military and nonmilitary means, such as intelligence, diplomacy, and humanitarian assistance, to advance U.S. national-security interests. Moreover, exhibiting cultural awareness and sensitivity vis-à-vis non-DoD partners is paramount to successful operational planning and execution. To build or bolster local governance, foster economic growth, and respond to natural disasters, the United States must use different types of tools, military and otherwise, simultaneously. It is no small task to synchronize these different tools so that they work in tandem, or at least do not conflict with one another.

There are a number of obstacles to increasing the effectiveness of integrated-operations training. These include a lack of qualified subject-matter experts, the inability or unwillingness of partner organizations to support integrated-operations training programs, and a tendency to focus on familiarization rather than in-depth understanding of non-DoD partners. While there are efforts under way to increase DoD integrated-operations preparation, which we detail in this book, many are hampered by budget constraints, limited staff, and uncertain prospects for their future existence. The military as a collective also tends to resist integrated-operations preparation, because time spent on the subject detracts from time that could be spent on more-traditional warfighting training.

Another major obstacle is the sheer complexity of the problem. Several different types of organizations are responsible for training the U.S. military. Each service has its own preferences for how it trains its people. In addition, combatant commands (COCOMs) and the Joint Staff also sponsor training and exercises. To the uninitiated outsider, the military-training community is highly chaotic, as it includes individual and unit training, training for different specialties and grades, and familiarization, continuity, and mission-rehearsal (MRX) training. In a similar way, the IA and international partners with which DoD must work represent another chaotic realm in terms of the number and variety of organizations involved. Training to work with these partners, then, involves the intersection or, some might say, the collision, of two complex communities.

To assist in overcoming these obstacles and to improve U.S. military capability, this study was intended to help Office of the Secretary of Defense (OSD) planners think about integrated operations and help them to craft a strategy to ensure that the U.S. military is better prepared to operate with nonmilitary and foreign partners in the future.

A New Approach to Preparing for Integrated Operations

The RAND study team found that almost all of the requirements for integrated-operations training can be found in existing joint and service task lists. From these lists, we derive a list of integrated-operations tasks, rank the tasks in terms of importance and training contribution, and then survey the training program to determine gaps. We argue that current training programs aimed at headquarters (HQ) staffs need to be revamped to focus on high-priority tasks that are amenable to training.

Recommendations

This book recommends that the Office of the Under Secretary of Defense (OUSD) for Personnel and Readiness (P&R) pursue a comprehensive approach to integrated operations that includes understanding such operations' component tasks, training requirements, and personnel management. Specifically, it recommends that OUSD P&R (1) conduct a demand analysis to determine the full nature and scope of DoD's integrated-operations capability requirements. Such an analysis would help ascertain how many integrated operators of what grades and specialties the U.S. military needs. This is a necessary step to developing a long-term solution. In the interim, DoD should (2) formalize an integrated-operations task list to encourage discussion between COCOMs, the services, the Joint Staff, and others in the training community. (3) In terms of training, this book provides recommendations on where to increase or decrease emphasis in an effort to make the best use of scarce resources. It recommends that OUSD P&R (4) create and maintain a database of ongoing training, exercises, and professional military-education courses, develop measures of effectiveness to analyze tasks and their application over time, and seek to gain insights into training activities undertaken by key allies that might be transferable to the U.S. military. The lack of visibility into training programs places significant limits on OSD's ability to oversee training efforts. OSD requires better information on the method and content of ongoing training programs to craft an effective plan for the future. Finally, once OUSD P&R has determined the most successful approaches, it should (5) advocate for stable funding for innovative programs, some of which currently exist but face uncertain prospects.

Acknowledgments

The authors owe a great debt to a number of military officers, civil servants, and analysts for their assistance on this study. These include people from OUSD P&R, the U.S. Joint Forces Command (USJFCOM) Directorate for Operational Plans and Joint Force Development (J7), USJFCOM Directorate for Strategy and Policy (J5), USJFCOM Joint Training Directorate and Joint Warfighting Center (J7/JWFC), U.S. Pacific Command Joint Interagency Coordination Group, the Marine Air-Ground Task Force (MAGTF) staff-training program, the Battle Command Staff Training Program, the U.S. Department of State (DOS) Office of the Coordinator for Reconstruction and Stabilization, the Joint Center for International Security Force Assistance (JCISFA), the U.S. Army–U.S. Marine Corps (USMC) Counterinsurgency Center Joint Assessment and Enabling Capability, the Joint Center for Operational Analysis, the USMC Center for Advanced Operational Culture Learning, and the Marine Corps University.

Within RAND, Forrest E. Morgan participated in some early work that set the stage for the rest of the project. Russell W. Glenn provided an extremely thorough review that helped strengthen the book considerably. We are also grateful for the insights provided by our external reviewer, COL Neale Cosby. Christopher Paul offered sage advice at several key moments. James Dobbins and Michael J. Lostumbo oversaw the work and generously offered their views.

The project officer for this study was Frank DiGiovanni, director, Joint Training and Ranges, OUSD P&R. Frank DiGiovanni, Joseph Thome, and Carlton Rosengrant provided outstanding support to the study on both substantive and administrative matters. We are grateful for their guidance and help throughout this one-year effort.

Abbreviations

ABN DIV	airborne division
AMETL	agency mission-essential task list
AO	area of operation
ART	article of the Army Universal Task List
AUTL	Army Universal Task List
BCTP	Battle Command Training Program
C2	command and control
C4ISR	command, control, communication, computer, intelligence, surveillance, and reconnaissance
CBRNE	chemical, biological, radiological, nuclear, and explosive
CI	counterintelligence
CJCS	chair of the Joint Chiefs of Staff
CJOA	combined joint operation area
CM	consequence management
CMO	civil-military operations
CMOC	civil-military operation center
COCOM	combatant command

COIN	counterinsurgency
COMMZ	communication zone
DCM	deputy chief of mission
DFID	Department for International Development
DHS	U.S. Department of Homeland Security
DOC	Directorate of Operational Capabilities
DoD	U.S. Department of Defense
DOE	U.S. Department of Energy
DOS	U.S. Department of State
EPW	enemy prisoner of war
FAO	foreign-affairs officer
FEMA	Federal Emergency Management Agency
FID	foreign internal defense
FSI	Foreign Service Institute
FSO	foreign-service officer
GCC	geographical combatant commander
HA	humanitarian assistance
HN	host nation
HQ	headquarters
HSC	Homeland Security Council
HUMINT	human intelligence
IA	interagency
IAW	in accordance with
ICS	Interagency Coordination Symposium

ID	infantry division
IO	information operations
IR	intelligence requirement
ISR	intelligence, surveillance, and reconnaissance
ITEA	Interagency Transformation, Education, and Analysis
J5	Directorate for Strategy and Policy
J7	Directorate for Operational Plans and Joint Force Development
J7/JWFC	Joint Training Directorate and Joint Warfighting Center
JIACG	Joint Interagency Coordination Group
JIATF	Joint Interagency Task Force
JIMPC	Joint, Interagency, and Multinational Planners' course
JMETL	joint mission-essential task list
JOA	joint-operations area
JSOU	Joint Special Operations University
JTF	joint task force
LEA	law-enforcement agency
LMTC	Leadership and Management Training Continuum
MCTL	Marine Corps Task List
MEF	Marine expeditionary force
METL	mission-essential task list
MiTT	Military Transition Team

MNE 5	Multinational Experiment 5
MOD	UK Ministry of Defence
MOE	measure of effectiveness
MOOTW	military operations other than war
MRX	mission-rehearsal exercise
MSTP	Marine Air-Ground staff-training program
MTN DIV	mountain division
NCA	national command authority
NCO	noncommissioned officer
NDU	National Defense University
NGO	nongovernmental organization
NSC	National Security Council
NTTL	Navy Tactical Task List
NTA	article of the Universal Naval Task List
OEF	Operation Enduring Freedom
OSD	Office of the Secretary of Defense
OUSD	Office of the Under Secretary of Defense
P&R	Personnel and Readiness
PA	public affairs
PDD	presidential decision directive
PIAP	police-information assessment process
PIR	priority intelligence requirement
PO	petty officer
POI	plan of instruction

PSYOP	psychological operations
PVO	private voluntary organization
S/CRS	Office of the Coordinator for Reconstruction and Stabilization
SCETC	Security, Cooperation, Education, and Training Center
SETAF	Southern European Task Force
SFS	senior foreign service
SOF	special operations forces
SOFIACC	Special Operations Forces–Interagency Collaboration course
SOS	special-operations squadron
SSTR	stability, security, transition, and reconstruction
T2	Training Transformation
TCCSE	Training Continuum for Civil Service Employees
TCFSG	Training Continuum for Foreign Service Generalists
TIM	toxic industrial material
TIO	tactical information operations
TRSS	terrorism-response senior seminar
TT	transition team
UE	Unified Endeavor
UJTL	Universal Joint Task List
UNTL	Universal Naval Task List
USAF	U.S. Air Force
USAID	U.S. Agency for International Development

USEUCOM	U.S. European Command
USG	U.S. government
USJFCOM	U.S. Joint Forces Command
USMC	U.S. Marine Corps
WMD	weapons of mass destruction

CHAPTER ONE

Introduction

The nature of recent challenges and the types of missions that the U.S. Department of Defense (DoD) has faced dictate that the U.S. military work with other government agencies, international organizations, private and nongovernmental organizations, and foreign militaries. To build or bolster local governance, foster economic growth, and respond to natural disasters, the United States uses a variety of tools, military and otherwise, simultaneously. It is no small task to synchronize these efforts so that they work together toward a common purpose.

This book addresses how the U.S. military can help prepare its personnel to work successfully with these partners (i.e., improve its *integrated operations*). Integrated operations are defined by those with whom DoD works, rather than by the type or location of an operation. In this sense, they are best thought of as an enabler for other activities, rather than a mission set of their own.

The partners with which DoD works in an operational context vary. Some are other U.S. government (USG) agencies, such as the U.S. Department of State (DOS), the Central Intelligence Agency, and the U.S. Department of Energy (DOE). This book refers to these as *interagency (IA) partners*.[1] Some are government agencies in other countries, such as interior and public-works ministries. We refer to them

---

[1] The DoD dictionary (USJCS, 2008, p. 270) defines *interagency* as "United States Government agencies and departments, including the Department of Defense. See also interagency coordination."

as *multinational partners*.[2] Some are militaries from other countries, which can be military forces from a host nation (HN) or militaries from other countries that cooperate with the United States in a particular operation. For this group, we use the term *coalition partners*.[3] General Richard Myers, the former chair of the U.S. Joint Chiefs of Staff, emphasized the need for coordination by bundling these three groups together under the term *integrated operations*.[4]

Garnering a full assessment of U.S. preparation for integrated operations requires assessing U.S. military training, exercise, and educational programs that support a variety of missions. Stability, security, transition, and reconstruction (SSTR) operations, for example, require the military to work with nonmilitary partners to gain and maintain order, provide social services, and encourage economic growth, all toward the end of rebuilding a community. Integrated-operations capability also enables major combat operations by building stronger coalitions. In short, integrated operations span the range of military operations.

U.S. military efforts to prepare for integrated operations bring together a number of players. Multiple actors play a role in training, educating, and exercising the joint force. The services play a major role via the execution of their basic Title 10 responsibilities, which include conducting basic training and continuing individual and unit training. The services maintain dedicated training venues, such as the U.S. Army's National Training Center and the U.S. Air Force's (USAF's) Red Flag program at Nellis Air Force Base. The Joint Staff and the geographical combatant commanders (GCCs) sponsor exercises, while the functional combatant commanders, such as U.S. Joint Forces Command (USJFCOM), commanders support the services and GCCs and offer their own training and exercises.

---

[2] The DoD dictionary (USJCS, 2008, p. 357) defines *multinational* as "Between two or more forces or agencies of two or more nations or coalition partners. See also alliance; coalition."

[3] The DoD dictionary (USJCS, 2008, p. 270) defines a *coalition* as "An ad hoc arrangement between two or more nations for common action. See also alliance; multinational."

[4] Myers (2005, p. 10). For more discussion of the definition, see Downie (2005).

The current recognition of integrated operations capability as an important attribute of the U.S. military force can be compared to the need for joint specialists. The experiences of Operation Eagle Claw (the ill-fated attempt to rescue U.S. hostages in Iran in 1980) and Operation Urgent Fury (Granada in 1983) contributed to the passage of the Goldwater-Nichols Department of Defense Reorganization Act of 1986 (P.L. 99-433) that significantly reformed Title 10. Goldwater-Nichols instituted reforms that have resulted in better cooperation among the military services and has transformed the armed services into a more joint military (see Locher, 2002). For example, officers must serve in a joint billet to be considered for promotion to general or flag officer. In a similar way, lessons learned from Operation Restore Hope (Somalia in 1992) and Operation Restore Democracy (Haiti in 1994) led to presidential decision directive (PDD) 56, Managing Complex Contingency Operations (Clinton, 1997), which was released in May 1997 (NDU, 2003; Murdock, 2004, p. 60). PDD-56 sought to improve IA training and planning in the area of peace operations.[5] Complaints about the lack of implementation of PDD-56, along with IA difficulties in Operation Iraqi Freedom, led PDD-56 to be superseded by National Security Presidential Directive 44 (Bush, 2005), which was signed in December 2005. Lessons learned from other military operations in Bosnia, Kosovo, and Afghanistan also highlighted the difficulties posed by working with coalition partners.

Study Objectives

This book seeks to assist DoD in its effort to increase its proficiency at integrated operations. We consider the types of tasks necessary to conduct integrated operations; assess how well the current slate of training, exercises, and education prepares DoD personnel to work with others; and make recommendations about how DoD might improve its preparation for these types of operations.

---

[5] FAS (1997) seeks to explain the directive.

Scope

Two considerations help to focus our examination of integrated-operations training. The first relates to operational context. Though integrated operations apply to a wide range of military operations, we selected SSTR operations for illustrative purposes. Given its current prominence, and the high degree of IA and international cooperation it requires, we believe SSTR operations to be the most representative mission to study integrated operations. Therefore, we do not detail the ways in which integrated-operations tasks differ across the broad range of military operations, nor do we attempt to assess the extent to which lessons relevant to SSTR operations are applicable to other types of operations.

Next, in our gap analysis, we limited our effort in terms of the type of military unit we surveyed. Given the size and breadth of U.S. military activity, our team focused its analysis on headquarters (HQ) staff. We chose these staff because they frequently interact with integrated-operation partners and because the data on staff training were the most consistently available. However, to provide additional insight and a fuller picture, we also examined, albeit to a less analytically rigorous standard, training for military advisers and other activities that help the U.S. military prepare for integrated operations.

Organization of This Book

First, we seek to understand the requirements for integrated operations. What are the necessary tasks to master to work successfully with IA, multinational, and coalition partners? Chapter Two describes how the team developed a list of these tasks, along with six broad categories to organize them. The chapter concludes by considering the relative importance of each task for mission success and by discussing to what degree training can contribute to the ability to perform each task.

Chapter Three describes the training activities reviewed by the team to determine where gaps exist in DoD efforts. To construct as complete as possible a set of cases, we focused on DoD efforts to train

HQ staffs. We also considered DoD efforts to train military-adviser units and other groups that also interact with integrated-operations partners. This training informed our thinking, but we separated these observations from the HQ-staff cases to preserve analytical clarity and rigor. Additionally, Chapter Three presents insights gained from discussions with IA, multinational, and coalition partners.

Chapter Four lays out the results of the task-list analysis and describes the several elements that should be considered when developing a strategy for conducting integrated-operations training. The chapter concludes with a discussion of the challenges that must be overcome in implementing such a strategy. Collectively, this chapter sets the stage for making some recommendations for how DoD can improve its efforts to prepare its personnel to work with others. Chapter Five considers and presents our overall conclusions and recommendations.

CHAPTER TWO

Identifying Key Integrated-Operations Tasks

Approach

After defining *integrated operations* and associated terms, the study team compiled a list of essential tasks for integrated operations. This chapter discusses how we developed our list. It also shows how we developed categories of tasks to help make them more understandable. Last, it shows how we prioritized the tasks based on their contribution to SSTR operations and according to the extent to which training contributes to "learning" how to achieve them. This approach provides a repeatable and traceable method to conceptualize the requirements for conducting integrated operations.

Why compile a list? U.S. military training relies on lists of "essential tasks" to develop its training objectives. To do this, unit commanders, seeking to fulfill the operational needs of the combatant commands (COCOMs), select a set of prescribed tasks from a larger, more general master list. These tasks have associated objectives that the unit seeks to meet in its training. The unit's list of tasks provides concrete goals to help orient and organize its training program.

With this in mind, we discussed integrated-operations training with various stakeholders, including service and joint training organizations, coalition-partner militaries, and officers from other USG agencies. Through these discussions, we gathered a variety of perspectives as to what types of tasks might be included in an integrated-operations task list. These discussions established the baseline for the subsequent research and analysis.

We began our research by surveying the Universal Joint Task List (UJTL) (USJCS, 2005a) and associated service task lists, including the Army Universal Task List (AUTL) (Headquarters, U.S. Department of the Army, 2003), the Marine Corps Task List (MCTL) (USMC, 2008), and the Navy Tactical Task List (NTTL) (NWDC, 2008). The USAF has recently adopted the UJTL in full, as opposed to developing its own service-specific task list.[1] The UJTL seeks to provide "a standardized tool for describing requirements for planning, conducting, evaluating, and assessing joint and multinational training" (USJCS, 2005a, p. 1). As such, it contains a comprehensive list of tasks that the armed services may be required to conduct in the course of planning, coordinating, and executing military operations. When preparing units for deployment, it has become increasingly common for unit commanders, in conjunction with other senior officers, to construct a mission-essential task list (METL) by selecting tasks from their respective service task list and the UJTL. Ideally, commanders should construct a training program designed to prepare their units to fulfill the tasks in the METL. In its most sophisticated form, each task is associated with a set of performance metrics (conditions and standards) that determine whether and to what level a unit is capable of performing the task. Having the relevant tasks in the METL can assist commanders in determining whether their units can fulfill their stated mission. Conversely, poorly conceived tasks are problematic, as they can impede a unit's efforts to prepare for the mission at hand.[2] This practice of developing METLs is not entirely followed in all services, nor by joint forces. This is due to a number of factors, from the different nature of operations in different domains (land, air, and sea) to the commander-

---

[1]   The U.S. Navy has done the same in the form of the Universal Naval Task List (UNTL) (NWDC, 2007).

[2]   The practice of using a task list with associated metrics is not free from criticism. For example, some have argued that DoD task lists frame functions that are not associated with particular types of missions and are fixated on metrics and standards that are only indirectly related to the ability to accomplish objectives. There are merits in these arguments, but they fall outside the scope of this effort, which takes the current training system more or less as given and offers suggestions for how to work within it.

centric nature of the training process, which favors the preferences of individual commanders over standardization.

Reviewing the Task Lists

The study team reviewed the task lists to select tasks that were within the unofficial Office of the Secretary of Defense (OSD) definition of *integrated operations* (from a draft DoD directive). The team cast a wide net when selecting these tasks, seeking to err on the side of inclusion, as it is easier to delete tasks from this list than to add tasks later. There is an element of human judgment in the study team's approach, but one advantage lies in its openness and traceability.

From the various lists, we identified 255 tasks relevant to integrated operations. We found that the UJTL contains 125 integrated-operations tasks, while the AUTL has 98, the NTTL 17, and the MCTL 15. Because the AUTL is more comprehensive than its service counterparts' lists, the U.S. Army has already incorporated a large number of integrated-operations tasks into its training structure. The U.S. Navy and U.S. Marine Corps (USMC) have not done so, perhaps because of their unique mission focus.

Challenges

There are some problematic aspects of tasks and how the team assessed existing training programs. Significantly, meetings with Army officials and private observers indicate that the METL-development process in the Army has become less disciplined and widespread in its employment in recent years.[3] For example, the examination of training objectives for the Army's Battle Command Training Program (BCTP) (which, in principle, should consist of a unit's METL) was hampered by the fact that records of the supporting tasks for training objectives were not readily available. There are often last-minute changes to training objectives in the lead up to actual BCTP exercises. Indeed, it has been reported that, in some mission-rehearsal exercises (MRX), train-

---

[3] For the Marine Air-Ground Task Force (MAGTF) staff-training program, some of its exercise data included not only training objectives but supporting UJTL tasks. This rarely happens.

ing objectives have not been finalized until the initiation of the actual exercise. Moreover, different training audiences address training objectives in different ways.

While impressionistic, it would appear that, without considerable additional research, there is no systematic manner by which the Army, and possibly the other military departments, can demonstrate a consistent trace between unit METLs and supporting training objectives. Moreover, this applies as well to demonstrating alignment with national-level policy guidance. As such, an examination of how well the military departments prepare their forces for integrated operations using DoD task lists provides an incomplete picture. However, there are sufficient data to demonstrate shortcomings in training for integrated operations and to illuminate problems in how the military departments set training priorities for their forces.

The Integrated-Operations Task List

Table 2.1 shows some illustrative tasks, drawn in this instance from the UJTL, that we deemed to be applicable to performing integrated operations. The entire list can be found in Appendix A of this book.

The 255 tasks address what the U.S. military needs to be able to undertake when working with IA, multinational, and coalition partners. In numerous discussions with veterans of integrated operations, representatives of the U.S. military and its partner organizations, scholars, and others, we found only one type of task relevant to integrated operations that is not present in the list: the need to transition organizations and operations from military to nonmilitary control.[4] Thus, the study team believes that the list serves as a sufficient basis to develop a program to prepare the military to work with non-DoD partners.

To make the list manageable, we divided the tasks into the six categories shown in Table 2.2. We did this by grouping similar tasks together by what we viewed to be major distinctions among them. Although there are cases in which some tasks could belong to more than one category, the team placed each task into the category where it fit best.

---

[4] We are grateful to Frank DiGiovanni for this point.

Table 2.1
Sample Tasks from the UJTL

UJTL Number	Joint Task Description	Abridged Joint Task Definition
OP 1	Conduct operational movement and maneuver	To dispose joint or multinational forces (or both); to affect the conduct of a campaign or major operation, either by securing positional advantages before battle is joined or by exploiting tactical success to achieve operational or strategic results
OP 1.2.4.8	Conduct unconventional warfare in the joint-operations area (JOA)	To conduct military and paramilitary operations, normally of long duration, within the JOA; includes, when appropriate, integration and synchronization of indigenous and surrogate forces
OP 1.5	Control operationally significant areas	To control areas of the JOA whose possession or command provides either side an operational advantage or denies it to the enemy; in military operations other than war (MOOTW), also pertains to assisting a friendly country in populace and resource control
OP 1.5.5	Assist HN in populace and resource control	To assist HN governments to retain control over their major population centers, thus precluding complicating problems that may hinder accomplishment of USJFCOM's mission
OP 2	Provide operational intelligence, surveillance, and reconnaissance (ISR)	To produce the intelligence required to accomplish objectives within a JOA; also includes intelligence support to friendly nations and groups
OP 2.1	Direct operational intelligence activities	To assist USJFCOM in determining its intelligence requirements (IRs), then planning the operational collection effort and issuing the necessary orders and requests to intelligence organizations; includes intelligence support to U.S. and HN forces in MOOTW
OP 2.1.1	Determine and prioritize operational priority intelligence requirements (PIRs)	To assist USJFCOM in determining and prioritizing its PIRs; in MOOTW, includes helping and training HNs to determine their IRs, such as in counterinsurgency (COIN) operations

SOURCE: USJCS (2005a).

Table 2.2
Integrated-Operations Task Categories

Category	Task
Establish relationships with partners	Create and promote relations between DoD and partners.
Provide security cooperation	Work directly with partners to promote security, including providing training and equipment.
Understand partner capabilities	Determine DoD and partner capabilities, and develop plan to use them to best effect.
Conduct operations with and for partners	Deploy, sustain, and extract; provide transportation and logistical support to partners.
	Protect partners; defend partners from attack.
	Provide consequence management; assist partners in recovering from catastrophe.
Collect and disseminate information	Collect information from partners and disseminate to joint force.
	Collect information from joint force and disseminate it to partners.
Support interpartner communications	Provide HQ or other elements that allow DoD and partners to coordinate their actions.

The first category, "establish relationships with partners," speaks to the need to create a basis for cooperation between DoD and its collaborators. In many cases, DoD is directed to work with particular partners and does not have the luxury of choice, but constructing strong relationships with others can help make the partnerships more effective, and possibly more efficient, than they would otherwise be. To the extent possible, if DoD can build good relationships with partners prior to a conflict, it will be much more effective when it comes time to work with them during operations. Direct contact and repeated interactions often help to build trust. Familiarity and knowledge are as important with respect to IA partners as to international ones. Our task-list survey found 15 tasks that fit into this category.

The second category, "provide security cooperation," includes efforts to provide equipment and training to partners.[5] These activities help increase partner preparedness and have the added benefit of improving DoD-partner relations. Theater-security cooperation, such as train, equip, advise, and assist programs, can promote stability and can actually prevent the need for an operation. DoD-provided assistance is usually associated with foreign partners, but the department can, and does, provide training and other help to state and local authorities and for employees of other government agencies. Our survey found 44 of these tasks.

The third category is "understand partner capabilities." Appreciating partner perspectives helps DoD to assess the limitations and strengths of the cooperative relationship. This task touches on an important aspect of cooperative relationships: realizing the comparative advantage. To the extent that DoD and its partners are able to capitalize on each other's relative strengths, the more effective they will be. Often, DoD is tasked to work with a particular partner to provide political or symbolic support for an activity. For example, DOS lists a number of the countries that have joined U.S. efforts in the war on terrorism (DoD, 2002a). However, in many cases, partners possess valuable capabilities that DoD lacks or are able to provide capabilities that allow DoD to apply resources from one area to another. Knowing that it can rely on German chemical-decontamination units or DOS-led reconstruction teams would help DoD reallocate assets elsewhere, resulting in an increase in operational effectiveness. We found 45 of these tasks.

The fourth category consists of tasks in which DoD "conducts operations with and for partners." This category incorporates DoD support to its partners that relate directly to an ongoing operation. It includes efforts to provide logistical support and transport to partners. It also consists of efforts to provide force protection for partners, some

---

[5]   DoD provides the following definition of *security cooperation*: "All Department of Defense interactions with foreign defense establishments to build defense relationships that promote specific US security interests, develop allied and friendly military capabilities for self-defense and multinational operations, and provide US forces with peacetime and contingency access to a host nation" (USJCS, 2008, p. 482).

of which do not have the means to protect themselves. Last, the category includes consequence management, which speaks to DoD capabilities to respond to natural disasters, or military attacks, that wreak widespread damage to populations. These are more or less "normal" military tasks, but we selected them because they would be conducted in an integrated-operation environment. There are 77 tasks in this category, making it the category that contains the most tasks.

The fifth category deals with efforts to collect and disseminate information. One advantage of working with others is that they often have access to different sources of information from DoD's sources. Obtaining access to that information can be quite useful. In addition, DoD needs to be able to share information with its partners to help ensure unity of effort. This category deals with more than intelligence; it is also about conveying and receiving information—to convey commander's intent, craft a public-affairs (PA) strategy, and ensure that DoD and non-DoD decisionmakers obtain the information they need to make informed choices. This category contains 65 tasks.

The final category we derived calls for DoD to "support interpartner communications." It is similar to the fifth category in that both deal with the ability to transmit information. It differs, however, in that it covers DoD efforts to provide the means for partners to coordinate their actions. In military terms, this category is about providing command-and-control (C2) capabilities. DoD should not assume that it will be the lead agency or even that the United States will be the lead country in every operation. However, whether in a supported or supporting role, it still needs to be prepared to ensure that it is able to create and sustain HQ elements that can organize forces for action, link different organizations, and provide a structure that allows leaders to react effectively to changing circumstances. There are nine of these tasks, making it our smallest category.

Setting Priorities

To allocate limited resources efficiently, it is essential to set priorities. Our team reviewed the 255 tasks and considered the impact of each on the success of an SSTR operation. To illuminate differences, we used a three-part scale: high importance, medium importance, and low or no importance. Highly important tasks were deemed critical to mission success. By *critical*, we mean that the mission would likely fail if the military were unable to perform this particular task. Medium-importance tasks would contribute significantly to mission success, and low-importance tasks would be nice to do but would not have much impact on the mission.

We used a modified Delphi approach.[6] Two team members independently reviewed the list and came up with rankings. These two met with a third team member and came up with a composite score. One team member is a retired Army colonel with extensive experience in operations research. Another is a veteran of the Strategic Studies Institute of the U.S. Army War College who currently works at the Naval Postgraduate School. The third has considerable experience working on requirements for coalition military operations. All three have Ph.D.'s in disciplines relevant to military operations and have considerable experience in studying military operations.

After coming up with a "team" score, we brought in an outside reviewer, a retired Army officer with experience as a planner in Afghanistan and a Ph.D. in military history. The outside reviewer made his own rankings, then met with the team to reconcile differences.

For each task, one team member assigned a priority, which was then discussed with at least two other team members. Again, this approach requires human judgment, and others might disagree with our results. We have documented the ranking for each task, making the process transparent and repeatable. We then summed up the results by task category. A summary of the rankings is shown in Figure 2.1.

If we rank the tasks by their importance to SSTR operations, our analysis suggests that "conduct operations with and for partners" ranks lowest. We found that 25 percent of the tasks associated with this type of integrated operations were high priority. The category of tasks that deal with providing security cooperation has the largest proportion of critical tasks of all six categories (59 percent).

---

[6] For more information on Delphi approaches, see Brown (1968) and Wong (2003).

Figure 2.1
Distribution of Tasks Within Skill Category, by Prioritization for SSTR Operations

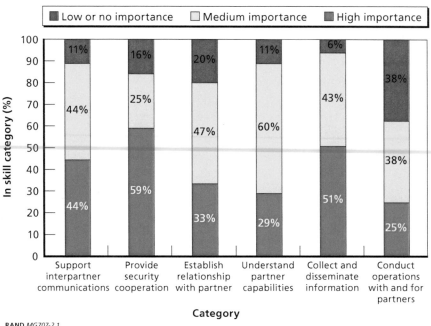

RAND MG707-2.1

Training Contribution

We also examined each of the tasks to understand whether proficiency in them is best gained through experience or training. Some tasks are simply more "trainable" than others. Research has shown that the strongest correlation between experience and job performance is seen when measured at the task level and when the measurement mode is the number of times performing the task (Quinones, Ford, and Teachout, 1995, p. 891). In the absence of deep experience, task repetition correlates with job performance. As a substitute for direct experience in performing integrated-operations tasks, training (and especially exercises) can have a significant impact on job performance. Moreover, the effect of gained experience on job performance is greatest for people with less experience. Additional experience does not improve performance as much for those with already high levels of experience. The utility in understanding which tasks are more trainable is that it can help identify where to focus subsequent training efforts.

Our assessment indicates that the category of establishing relationships with partners can be critically affected by a training intervention of some form. In fact, as shown in Figure 2.2, all categories can indeed be affected to some degree by training. The tasks that we considered to be most amenable to training were in "establish relationship with partner" and "collect and disseminate information" categories. Even the lowest category in this scoring, "support interpartner communications," has a good proportion of tasks that we judged to be trainable.

Figure 2.2
Distribution of Tasks Within Skill Category, by Training Contribution

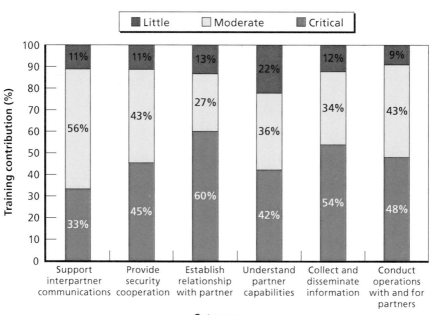

Conclusion

This chapter has shown how we derived a list of integrated-operations tasks, categorized the tasks, and considered the tasks in terms of their importance to SSTR operations and how well they can be trained. As mentioned earlier, Appendix A provides the entire list of integrated-operations tasks, along with the priority and training-contribution rankings. The final step before offering recommendations is to survey the slate of training opportunities available, to determine the existence of gaps in the training program. That survey is presented in Chapter Three.

CHAPTER THREE

Identifying Gaps in Training Activities

The previous chapter described the set of essential tasks for the integrated operator. In this chapter, we discuss how well the current slate of training and exercise events promotes these skills. To do this, the study team surveyed several training and exercise programs that incorporate integrated-operations tasks. In all, we examined 10 separate programs. Currently, there simply is no DoD "one-stop shop" for preparing integrated operators. Instead, different integrated-operations skills are taught in different venues.

After comparing our findings on the gaps in the training program with our findings on the criticality of the tasks and their training contributions, we will have a more accurate view of the current level of integrated-operation training that takes place in DoD.

Approach

The study team examined several training and exercise programs that have at least a partial focus on developing integrated-operations skills. There are a number of ways in which the U.S. military prepares its personnel for operations. Training, for example, is an activity that provides specific skills that might be of use in an upcoming mission or for a finite period of time. DoD trains its personnel both as individuals and as units. Education, in contrast, is an activity that provides personnel with broader knowledge aimed at career-long development (USJCS, 2005b). DoD also prepares its people by having them participate in exercises designed to test various skills.

The team took two approaches to gain insight into important aspects of how U.S. military personnel prepare for integrated-operations activities. First, we looked at activities designed to prepare HQ staffs that are likely to interact with non-DoD partners. These staffs play a key role in forming relationships with integrated-operations partners and in planning operations with them. These interactions often have far-reaching effects on operations. Such staffs include COCOM and component command staffs. In many cases, staffs associated with corps, divisions, and brigades can make similar contributions.

Team members conducted focused discussions with course coordinators, instructors, students, and subject-matter experts serving on HQ training staffs in the U.S. Army, USMC, USJFCOM, and the Joint Staff. For joint training programs, the team reviewed the associated plans of instruction (POIs) and other related course materials to identify the integrated operations–related elements. For service-specific training programs, we examined unit training objectives. To better appreciate the content of the programs, team members observed actual classroom activities when possible.

The team also surveyed several training programs that focus on training U.S. troops to advise foreign militaries. This is certainly an important example of integrated operations, borne out by recent U.S. experience in Iraq and Afghanistan.

In the following section, we first provide a brief description of each activity observed and then describe which elements of the program relate to the integrated-operations skill set identified in Chapter Two. Next, the chapter provides a discussion of both the coverage of the integrated-operations tasks and the gaps for the programs focused on HQ staffs. These activities had the most consistent data in terms of training objectives and programs of instruction. The chapter then discusses our observations from military-adviser training and other programs. Finally, the chapter provides a brief discussion of some non-DoD programs for comparison.

Staff-Training Activities

The study team reviewed seven training activities aimed at HQ staffs: (1) the Interagency Transformation, Education, and Analysis (ITEA) program's Interagency Coordination Symposium (ICS), (2) the Joint Special Operations University (JSOU) Special Operations Forces (SOF)–Interagency Collaboration course (SOFIACC), (3) the JSOU terrorism-response senior seminar (TRSS), (4) the Joint Forces Staff College's Joint, Interagency, and Multinational Planners' course (JIMPC), (5) USJFCOM Joint Task Force (JTF) HQ training, (6) BCTP, and (7) the MAGTF staff-training program (MSTP).

Chapter Two identified and grouped the integrated-operations tasks into six broad categories. The training courses reviewed by the study team cover only a portion of those tasks. Moreover, in some cases, a task may be partially addressed by a course. For example, the ICS training only partially addresses task OP 5.7.2 ("determine national or agency capabilities and limitations"). The course focuses only on integration techniques with IA organizations and does not address integration techniques with multinational or coalition partners. Despite providing relevant, focused training, few programs observed by the team address working with IA, multinational, and coalition partners. This is not a shortcoming of the programs; rather, it simply reflects the focus of the individual training courses.

This section reviews how each of the programs relates to each of the six broad categories of integrated-operations skills. Within each category, the team linked specific tasks to courses or blocks of instruction within each of the programs. Tasks were considered to be widely taught if four or more of the programs addressed it and narrowly taught if three or fewer of the programs addressed it. Detailed analysis is contained in Appendixes B, C, and D. Appendix B provides an analysis of the first four courses, correlating the training objectives with specific tasks from the six task categories. This level of analysis permits a relatively close look at task coverage and provides insight into which tasks are currently being emphasized in courses that aim to provide training in integrated operations. This same convention is used for the discussion regarding military advisers (detailed analysis in Appendix E). The analyses in Appendixes C and D differ somewhat from that in Appendix E and are explained further in the following sections. What follows is an activity-by-activity review of the courses examined.

Interagency Coordination Symposium

The ITEA program at the National Defense University (NDU) conducts the ICS. The ICS is a strategic-level seminar, aimed at Joint Interagency Coordination Group (JIACG) representatives and Joint Staff officers. The seminar is held once per quarter in Washington, D.C., and can also be provided as a tailored, on-site version when requested by the COCOMs. The course normally runs two to three days on site and lasts for a full week when held at NDU.

ITEA ICS program coordinators believe that, in addition to providing valuable IA training, the program is helpful in informing the debate as to what kinds of skills an integrated operator should possess and what corresponding training curriculum should be developed.[1]

ITEA seminar presenters are typically brought in from various outside agencies and departments to bring relevant IA views and lessons from experts in the field. Individuals with JIACG experience are often called on to lecture at the seminars. Relevant blocks of instruction within the symposium include

- Interagency Coordination Overview: The National Security Council and the Homeland Security Council (HSC)
- Coordination with State and Local Governments
- Coordination Challenges Case Studies/Lessons Learned
- USG Capabilities and Coordination Exercise (ITEA, 2006).

Special Operations Forces–Interagency Collaboration Course

The SOFIACC is conducted at the O-4 level and below. The course was developed in 2006 and, at the time of this writing, has been taught only once, with three offerings planned for calendar year 2007. SOFIACC focuses only on preparing military personnel to work with the IA and

---

[1]   Discussion with ITEA program coordinators.

does not address working with multinational organizations or coalition partners. Relevant blocks of instruction include

- The National Security Council [NSC] and the Interagency Process
- The Interagency Process
- Collaboration with Other Agencies
- The Embassy Country Team
- The Joint Interagency Coordination Group (JIACG)/Joint Interagency Task Force (JIATF)
- Collaboration with Intelligence Agencies
- Shaping the Environment: Security Assistance and Foreign Internal Defense [FID]
- SOF-IA Collaboration Exercise (JSOU, 2006a).

Terrorism-Response Senior Seminar

The TRSS, developed in 1977, primarily targets special-operations personnel at the O-5 level and above. FBI and DOE representatives often attend, but there is typically no other IA participation, with the exception of the invited speakers. Class size is normally 25–30, with 15 as the minimum. Presentations are made by guest speakers from various organizations within the IA group. JSOU provides the course objectives to the speakers ahead of time to help them prepare and keep the discussions on track. Briefings last about 25–30 minutes, with about 45 minutes for questions and answers, allowing for student interaction with the speakers.

The course is offered several times per year; however, few participate. The study team had planned to observe a portion of the TRSS during a visit to JSOU, but the seminar was cancelled due to insufficient IA participation. When the course is held, relevant blocks of instruction include

- Interagency Roles
- Roles of Other USG Agencies
- Security Assistance and FID

- Terrorism Response Case Studies (JSOU, 2006b).

Joint, Interagency, and Multinational Planners' Course

The JIMPC is offered five times per year with a class size of about 25. It was taught for the first time in January 2006 and has so far been conducted only three times. The target audience for the course is the JIACG action officer, ideally with a mix of 50-percent DoD and 50-percent non-DoD participation. The non-DoD participants account for as much as 30 percent.[2] Blocks of instruction are taught primarily by guest lecturers, in much the same way as the ITEA and JSOU courses.

The focus of the course is on helping the student understand how NSC guidance affects COCOM planning. Relevant blocks of instruction include

- The Interagency Process
- Interagency Players in Complex Contingencies
- The Country Team
- The Ambassador/Country Team and Military
- The JIACG Concept
- Intelligence Support to the Interagency
- NGOs [nongovernmental organizations] and Transnational Corporations
- Interagency Exercise, "COHERENT KLUGE" (JFSC, 2006).

Joint Forces Command Training for Joint Task Force Headquarters

Recent reforms in transforming joint training have included placing importance on integrated operations. In accordance with the Unified Command Plan (DoD, 2002b), USJFCOM provides joint training expertise, inter alia, to the COCOMs and the services. This is achieved by supporting the development of joint training requirements and methods and the execution of joint exercises. A key element of the joint-exercise program is the series of exercises that USJFCOM orga-

---

[2]   Discussion with course coordinator, December 5, 2006. IA representatives are primarily country team members serving in U.S. embassies abroad.

nizes for each of the geographic COCOMs. Typically, USJFCOM will organize two to three MRXs each year for each of these commanders. In addition, as part of OSD's Training Transformation (T2) initiative, USJFCOM has become one of the key actors in bringing T2 approaches to joint training and exercises to JTFs. An essential tool in furthering T2 objectives and training for COCOMs and JTF HQs has been USJFCOM's singular mission of accreditation and certification of key capabilities of COCOM staffs. Thus, through a series of exercises organized by USJFCOM, COCOMs and JTF-HQ staffs are tested in critical areas identified as essential for mission success (USJCS, 2002, pp. G1–G8).

One of the principal tools employed by USJFCOM in supporting the joint training objectives of the COCOMs has been the Unified Endeavor (UE) exercise series. Begun in 1995, UE exercises are for JTF component commanders and their staffs to train at the operational level of war. Typically, UE exercises contain a three-phase program that ends in a simulation computer-aided exercise. The three exercise phases consist of academic training, the development of operational plans, and the execution of operational orders. For a full analysis of the training objectives of these exercises, see Appendix C. The analysis in Appendix C is presented in a tabular format, correlating each of the six task categories with the various exercises conducted by USJFCOM. Each set is color-coded to indicate the level of focus that a given task area was afforded.

Battle Command Training Program and the U.S. Marine Corps Marine Air-Ground Task Force Staff Training Program

It must be stated that the analysis of BCTP and MSTP cannot be examined in direct comparison with the findings of the training programs in the preceding section. The operations of both organizations are based on the principle of "task-organizing" themselves to meet units' immediate training requirements as established by their commanders prior to deployment. Neither program bases its training on formal POIs. As a result, the study team examined the training objectives of BCTP- and MSTP-sponsored training programs.

Notwithstanding the timely responsiveness of the management of both organizations to the RAND study team's entreaties for information, it has not been possible in the analysis of stated training objectives to examine specific "tasks" (either service- or UJTL-specific) in all instances. The ability to do so would have allowed the study team to examine with greater fidelity the degree to which specific integrated-operations tasks are trained. Indeed, if such training objectives were based on a strict usage of the UJTL in all training events, this would enable a disciplined cross-service and historical assessment of the mission-essential tasks that senior commanders are determining are necessary to meet envisaged operational requirements. Such a reform would have the additional benefit of enabling combatant commanders, the Joint Staff, and OSD to better understand and influence the training executed by the military departments.[3]

An additional analytical complication in this approach is that training objectives are comprised of one or more specific task(s). Finally, as mentioned in Chapter Two, meetings with U.S. Army officials and private observers confirm that the formal METL-development process in the Army has become less disciplined and more atomized in its employment and application in recent years, which prevents observers from examining common AUTL tasks in which units are being trained. More specifically, with respect to BCTP training, we observed the following:

- One will frequently see considerable last-minute changes to a unit's training objectives in the lead-up to a BCTP exercise. Indeed, it has been reported that, in some MRXs, training objectives have not been finalized until the initiation of the actual exercise.

---

[3] One can predict that such an initiative would likely encounter opposition from the military departments as an infringement on their responsibilities and functions as established in U.S. Code. Specifically, 10 USC §3013 specifies the key 12 functions of the U.S. Army: recruiting, organizing, supplying, equipping (including research and development), training, servicing, mobilizing, demobilizing, administering, maintenance, and infrastructure. They are the same for the other two military departments (§5013 for the Navy and §8013 for the USAF).

- One finds wide variety in the way different training audiences address training objectives. In general, there is some consistency within units but not always between units. This may well be a manifestation of how table of organization and equipment units have strayed from strictly using the AUTL.

Appendix D presents matrixes that present the study team's full analysis of the exercise-series data that BCTP and MSTP provided.[4] The analysis in Appendix D is given in a tabular format similar to that provided in Appendix C. Each of the six task categories is correlated to the various exercises conducted under the MSTP and BCTP programs. The same convention of color-coding is used to indicate the level of focus that a given task area was afforded.

A number of observations can be drawn from a review of these data with respect to how they address integrated skill sets:

- Establishing relations with partners in BCTP appears to have diminished as a critical training objective, whereas MSTP sees this as a continuing training priority. That said, the RAND study team has insufficient data to make any firm determinations in this regard, as it could well be that Army commanders hold the view that, after multiple deployments to Iraq and Afghanistan, such skills are already well understood by their units.
- Providing security cooperation would appear to be a critical area of concern. The data show that neither USMC nor the U.S. Army is consistently addressing these tasks in these exercises.
- Understanding partner capabilities appears to be an area of inconsistency at best and ambivalence at worst in both BCTP and MSTP. Again, there are insufficient data to judge whether such tasks are not deemed as essential by commanders.
- Conducting operations with, and for, partners is the one skill-set area in which both BCTP and MSTP excel. However, this otherwise positive observation should be tempered by the fact that

---

[4] The exercises I Corps, Yama Sakura 49 (January 2006), and 2nd Infantry Division (ID) (May 2006) are omitted from this immediate analysis, given that they were bilateral exercises, largely with theater-specific training objectives.

many of these tasks are basic warfighting tasks that both the U.S. Army and USMC should be expected to do very well.

- Collecting and disseminating information is another integrated-operations skill-set area in which both BCTP and MSTP have placed priority—the Warfighting Exercise for the 38ID, as recently as October 2006, being an unusual exception.
- Ensuring interpartner communication is yet another integrated-operations skill set that is almost universally covered by both programs, with one UE exercise in October 2005 being an unusual exception to this observation.

These data, in aggregate, offer some interesting observations. First, MSTP would appear, from the data obtained, to consistently cover integrated-operations skills to a greater degree than does BCTP, with the provision of security cooperation remaining a conspicuously glaring exception. Moreover, an examination of these data does not manifest any clear trends, i.e., demonstrable learning curves whereby integrated-operations skills are increasingly being vetted and validated by either MSTP or BCTP. Greater transparency and the consistent usage of such critical training tools, such as the UJTL or even the AUTL, would be welcome and could possibly disprove these observations.

In Figure 3.1, we present aggregate findings for how well the surveyed staff programs cover the task categories.

Military-Adviser Training Activities

Like the integrated-operations training being conducted for staffs, the military-adviser training activities offer insights into how U.S. military personnel are being prepared to work with and provide training and assistance to multinational and coalition partners. This section describes four such activities and provides an analysis of integrated-operations task coverage in Appendix E.[5] While these

---

[5] Only two are addressed in Appendix E: the Military Transition Team (MiTT) and 6th Special Operations Squadron (SOS) (6SOS). Data to complete a similar analysis for the

Figure 3.1
Coverage of Task Categories by Selected Staff-Training Activities

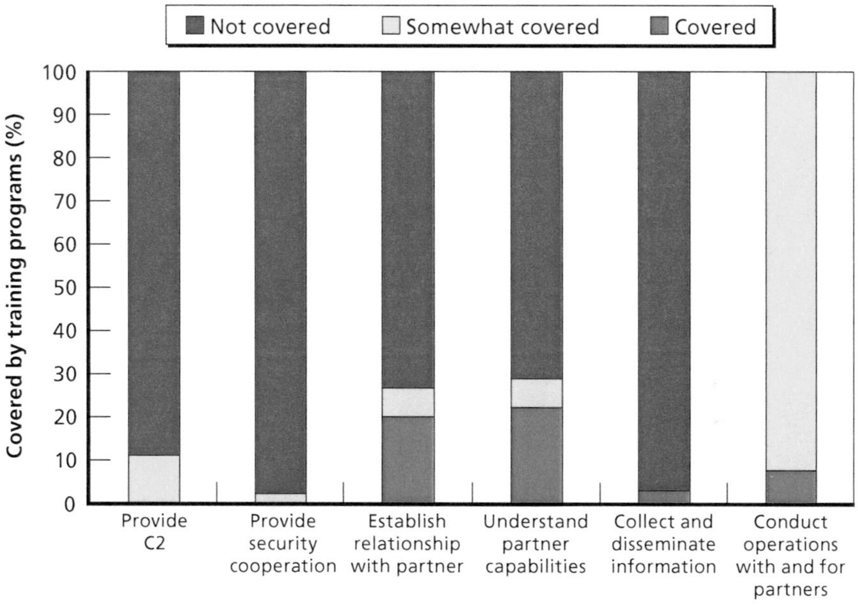

RAND MG707-3.1

activities place a great deal more emphasis on working with multi-national and coalition partners than the other activities do, they put very little on working with IA partners.

Military Transition Team Training

To prepare personnel for the task of training the incipient Iraqi military, the U.S. Army has retasked elements of 1ID to enable personnel and the organization to focus solely on MiTT training (see Spiegel, 2006). The 1ID at Fort Riley conducts training to prepare MiTTs for deployment to Iraq. MiTTs are often 10-person teams, primarily made up of conventional forces. The teams are assigned to the Iraqi Assistance Group and are embedded in Iraqi units, mostly at the battalion and

other two—Security, Cooperation, Education, and Training Center (SCETC) and COIN— were not available.

brigade levels. MiTTs at the brigade level typically have one special-operations officer (major or lieutenant colonel) assigned to them.

1ID focuses MiTT training largely on the tactical and technical skills the transition team (TT) will need to impart to its Iraqi counterparts. Few courses teach members how to effectively transfer that knowledge. Specifically, during the 54-day program, 1ID reserves just three afternoons for cultural awareness and three days for teach and advise training. It devotes the remaining 50 days to learning tactical skills. Several sessions on language training are provided but generally teach survival language skills only.[6] Officers involved in the training have identified other shortcomings. For example, in July 2006, one former battalion commander provided a written critique of the program, emphasizing the need for additional training in language skills, cultural awareness, negotiating, building rapport, and interpersonal skills.[7] According to the officer, MiTT members often fail to comprehend the vast cultural differences they will face. This cultural divide has led to some instances in which MiTT members have simply avoided interaction with their Iraqi counterparts. The obvious implication is that the mission is negatively affected. Specific courses related to the integrated-operations tasks identified in Chapter Two include

- Cultural Awareness
- Teach and Advise (Nagl, undated).

Security Cooperation Education and Training Center

Due to the high level of demand for TTs, USMC has recently created the SCETC to augment USMC's ability to provide this critical capability and to standardize training procedures. In the past, individual Marine expeditionary forces (MEFs) conducted this training on their own. In addition to providing MiTTs, USMC is providing border transition teams and national police-training teams, even though these are not normally USMC mission areas. Some of the training is conducted

---

[6] Discussion with U.S. Army lieutenant colonel assigned to 1ID, Fort Riley, Kansas, responsible for conduct of MiTT training.

[7] Discussion with U.S. Army major formerly assigned to a MiTT.

at units' home stations, and some is conducted at a newly constructed facility at Twentynine Palms in California. Unlike the Army's MiTTs, USMC personnel usually come from the same unit.

Like the Army's effort at Fort Riley, USMC focuses on the skills that it will need to teach to its foreign partners. SCETC also provides language and cultural training. It uses role-players and has marines act out scenarios, which help teach the TTs about how to work with people from other cultures. However, at the time of this writing, the training does not include interaction with other USG agencies, such as the Immigration and Customs Enforcement section of the U.S. Department of Homeland Security (DHS), which has worked with the Army's 1ID.

USMC is working to institutionalize and strengthen its ability to train military advisors. It is divesting SCETC of its training responsibilities and is working to create the Marine Corps Training and Advisory Group, an independent command that will organize, train, and equip adviser teams (MCTAG, 2007).

Combat-Aviation Advisor Mission-Qualification Course

6SOS at Hurlburt Field, Florida, conducts the Air Force Special Operations Command Combat Aviation Advisor Mission Qualification Course. The course seeks to prepare the squadron's personnel to serve as advisers and to provide training to foreign air forces. This is the squadron's primary mission.[8]

The course consists of four phases, and, like the TT training conducted by 1ID and SCETC, it teaches a mix of technical skills and integrated-operations skills. Through a combination of lecture and practical application, 6SOS ensures that its members are versed in cross-cultural communication and integration techniques, regional studies, instructor and adviser techniques, security-assistance management, and interpreter and translator operations. The culmination of the second phase of training is the weeklong Raven Claw exercise, in which the students apply approximately three months of academic training.

---

[8] Telephone discussion with 6SOS personnel, January 30, 2007.

The students are evaluated on their technical skills as well as their ability to apply the integrated-operations skills they have learned.

Following the Raven Claw exercise, the students complete two to four months of intensive language training before finally beginning the fourth phase, which consists entirely of technical aviation–related skills. Specific courses related to the integrated-operations tasks identified in Chapter Two include

- Methods of Instruction
- Advisor Techniques
- Security Assistance Management
- Political/Cultural Integration Techniques
- Civil-Military Operations
- Dynamics of International Terrorism
- Defense Security Assistance Management
- Contemporary Insurgent Warfare
- Regional Orientation
- Cross-Cultural Communications (AFSOC, undated).

U.S. Army–U.S. Marine Corps Counterinsurgency Center

The Counterinsurgency Center at Fort Leavenworth is a small but innovative program that seeks to change the U.S. military's mind-set to increase its capability to combat insurgents. Instead of focusing on tactics, it promotes the sort of critical thinking necessary to successfully conduct COIN. Staffed by both the U.S. Army and USMC, it stages weeklong seminars for MiTT teams from both services, mixing classroom seminars and panel discussions with previously deployed personnel. It emphasizes the need for cultural awareness, equating it with force protection and even mission accomplishment. Instructors also encourage a high level of interaction between TTs and their foreign counterparts.

Other Activities

In addition to the programs described already, the study team considered the contribution of other programs to integrated-operations training. Specifically, the team looked at national-level exercises that are designed to include IA and multinational partners. These exercises are designed to provide training on specific tasks, which often includes tasks related to integrated operations. Despite this, Joint Staff and USJFCOM officers responsible for national-level exercise programs generally indicated that exercises are not currently effective in training integrated operators.[9] One reason for this is the transient nature of the HQ staffs and other personnel who participate in the exercises. Participation is not related to a career continuum of training, but instead is based on who occupies a given position on a given day. As a result, exercises are not a reliable way to provide training over time, particularly if the goal of training is to provide a skilled cadre of integrated operators. The opinion commonly expressed was that a more formal, classroom-based integrated-operations training effort would build the foundation necessary for effective integrated-operations exercises. However, even if a robust integrated-operations training program were in place, there are other problems inherent in the national-level exercise program.

For example, there is currently a lack of IA participation in chair of the Joint Chiefs of Staff (CJCS) and COCOM exercises, despite the existence of a five-year plan that lays out national-level exercises for IA training. Outside of DoD, a common limitation is the availability of personnel to participate in exercises. In many cases, offices are "one-deep," and participating in an exercise means forgoing other, important work, i.e., opportunity costs. However, staffing levels in other organizations only partially explain the participation problem.

Another major problem with national-level exercises is the coordination of training objectives. National-level exercise training objectives within the five-year plan are coordinated and approved by the NSC's Plans, Training, Exercises, and Evaluations Policy Coordina-

---

[9] Discussion with Joint Staff Directorate for Operational Plans and Joint Force Development (J7) Exercise Branch officers, October 18, 2006, and USJFCOM Joint Training Directorate and Joint Warfighting Center (J7/JWFC) personnel, November 13, 2006.

tion Committee. However, as one Joint Staff colonel pointed out, "The interagency doesn't like to fail." This approach is antithetical to the military's understanding of exercises, in which the purpose of such endeavors is precisely to ascertain where there are points of failure and to use the entire process as a positive learning tool for participants. As a result, these national-level exercises are tightly scripted, and not every department's objectives are incorporated, limiting the ability to learn from mistakes.

A related problem, repeatedly voiced during discussions with experts, is the use of IA participants as "training aids" for the military. In other words, if a script calls for interacting with a DOS representative, it is preferable to have an actual DOS official with whom to interact. This can be boiled down to the issue of including each agency's training objectives in the script, not just DoD's training objectives. The problem, as expressed by one DOS official, is that DoD is not interested in learning about how DOS would actually respond to the scripted situation. He explained that, in one instance, despite participating in the scenario-planning meetings for exercises with foreign partners, DoD did not incorporate what he believed were realistic training objectives for his department. The result, as he put it, was that he would have to engage the foreign partners after the exercise to "straighten out the mess."[10]

In addition to exercises, the study team observed a USJFCOM experiment called Multinational Experiment 5 (MNE 5). MNE 5 is the fifth in a series of experiments conducted by a core group of multinational experimentation partners. Participants for MNE 5 include Australia, Canada, Denmark, Finland, France, Germany, Sweden, the United Kingdom, NATO's Allied Command Transformation, and the United States. In addition, U.S. IA participants include the DOS Office of the Coordinator for Reconstruction and Stabilization (S/CRS). MNE 5 is attempting to develop integrated planning, execution, and assessment capabilities among the participants and test certain concepts for potential employment in crisis-prevention and

---

[10] Telephone discussion with DOS Foreign Consequence Management office, November 3, 2006.

consequence-management activities. The military aspect of MNE 5 uses an effect-based approach to multinational operations as the contextual theme to facilitate military support within IA operations (USJFCOM, undated).

While experiments have their utility in bringing together agencies, multinational organizations, and coalition partners to essentially discuss concepts for integrated operations, it is not clear to what extent the good thinking that comes out in these discussions can be put into practice. Moreover, because some experiments do not have set deliverables or even concrete objectives to achieve in a given period, it is most difficult to measure the overall effectiveness of each event. Further, discussions with DOS and British, Danish, and Swedish contingents indicate that these entities are unsure of their future participation in USJFCOM experiments due to being one-deep at home.

Other Perspectives

The preceding discussion sheds considerable light on how DoD is currently approaching training for integrated-operations tasks, but the team also considered how other IA and foreign partners approach it. Within the IA, we found that there is often little emphasis on formal training. For example, the U.S. Agency for International Development (USAID) has no formal training for working with other departments, such as DoD. In fact, USAID most often hires experts with extensive backgrounds and experience in other departments.[11] This reliance on other departments to provide the necessary training seems to work well and is common in many of the smaller departments. DOS takes this approach as well, going as far as to explain in its Training Continuum for Civil Service Employees (TCCSE) that, "as a mid-level employee, you will generally have had several years of experience in the Department or as a federal employee in another agency." To gain more insight into non-DoD training, the study team conducted focused discussion

---

[11] Discussion with USAID Office of U.S. Foreign Disaster Assistance staff member, October 20, 2006.

with representatives from the DOS Foreign Service Institute (FSI). As one of the larger departments of the government and, moreover, one with which DoD personnel routinely interact, examining DOS's approach to integrated-operations development is appropriate. Moreover, the practices of some of key U.S. allies, such as the United Kingdom, are also considered.

The Foreign Service Institute

Within DOS, there is very little in the way of mandatory training. DOS provides two types of training: the Leadership and Management Continuum (LMTC) and the tradecraft courses.

For its tradecraft training, DOS uses TCCSE and the Training Continuum for Foreign Service Generalists (TCFSG) (DOS, undated[a], undated[b]). A subset of these is LMTC (DOS, 2004). While TCCSE and TCFSG courses are generally optional, as personnel progress from level to level within the organization, they are required to take certain courses from the LMTC. In general, the amount of emphasis on IA activities increases as personnel rise in seniority. A junior foreign-service officer (FSO), for example, will undergo tradecraft orientation training followed by mandatory basic leadership and management training. Later in his or her career, the FSO has the option to take additional tradecraft courses. He or she may also be required to complete additional leadership and management training, but these courses are taken only after the oficer is selected for promotion.[12]

To further illustrate the DOS approach, the team looked at the foreign-affairs officer (FAO) specialty. This is the largest single specialty within DOS, and FAOs are especially interesting, as they typically engage in integrated operations–type activities. FAOs are expected to lead IA coordination efforts for foreign-policy issues, as well as lead and coordinate U.S. representation abroad. Within the TCCSE, there are suggested and recommended courses that are intended to provide the FAO with these types of skills. For example, the mid-level FAO is encouraged to take such courses as Multilateral Diplomacy and to "go

---

[12] Discussion with DOS FSI faculty, November 30, 2006.

on informal visits (individually or with your supervisor or co-workers) to meet counterparts in other agencies/organizations."

At the senior level, FAOs are encouraged to take a rotational assignment within another USG agency. The practice of having FAOs serve in other government agencies helps considerably in facilitating IA cooperation. It is important to keep in mind, however, that these are only recommendations and suggestions, and the FAO is not required to complete any of the TCCSE development activities. In fact, these recommendations are not significantly different from the training recommendations for all DOS foreign-service employees. In essence, then, only the seniority-related, mandatory LMTC courses should be considered when looking at how DOS develops its officers to engage in integrated operations.

The mandatory LMTC courses begin at the junior level and progress through the senior foreign-service (SFS) and ambassadorial ranks. The first three mandatory courses (junior, mid-, and senior levels) are five days each in length and do not address integrated-operations skills. At the highest level (deputy chief of mission [DCM] and above), however, the training becomes significantly longer (three weeks) and takes on a different focus. The DCM course, for example, lists its first objective for participants as learning to "build teams across divisions and agencies at post," and the SFS course prepares the newly promoted officer "for the challenges they will face leading across agency and national boundaries."

Observations from Foreign Training: United Kingdom

For the purposes of comparing integrated-operations training practices with a foreign ally, it is worth highlighting the case of the UK. Two offices within the Ministry of Defence (MOD) are particularly interesting in that they have a responsibility for ensuring that the military is prepared for integrated operations. First, the Stabilisation Unit (created in 2004 as the Post Conflict Reconstruction Unit) is part of MOD but is funded by the Department for International Development (DFID) (similar to USAID). Unlike the U.S. system, DFID has virtually all of the national-level resources for postconflict operations (i.e., SSTR operations). The Stabilisation Unit includes both military and civilian officials who share responsibility for conflict assessment and planning. While the unit does not have a formal, doctrinal role, it does have the ability to influence UK military-training practices through its "comprehensive-approach" strategy, which seeks to bring military and civilian officials together to improve operational planning. The Stabilisation Unit fills gaps in training, such as developing concepts and hosting workshops on how to transition from a military to a civilian operational lead. It also offers joint operational-training courses on provisional reconstruction teams and includes civilian partners. However, it does not include foreign nationals in this training, which could be considered a shortfall. It is worth noting that the Stabilisation Unit is also a regular participant in USJFCOM's MNE 5 experiment, discussed earlier. However, unit officials explained that it is difficult to keep up this commitment to MNE 5 due to the small number of unit personnel.[13]

The second independent organization within MOD is the Department of Operational Capabilities (DOC), which has a rather unusual role as the central point for operational lessons identified.[14] DOC connects the operational lessons from the field directly to UK joint training programs, helping to ensure training relevancy.[15] DOC works closely with DFID, the Stabilisation Unit, the Foreign Office, and other UK civilian ministries. However, DOC's only significant foreign partner is USJFCOM. DOC focuses about half of its efforts on operational lessons and about half on feeding operational lessons into the collective training apparatus.

The overall philosophy of MOD is that the military will not "go it alone"—it very much relies on its partners within the civilian-dominated ministries, and, as a result, training together is second nature. The incorporation of civilians (Foreign Office and DFID) into one key military

---

[13] Discussions with MOD officials, London, UK, July 2006.

[14] MOD refers to *lessons identified* rather than *lessons learned*, which, in its opinion, occur only when an action is taken to correct a deficiency.

[15] As of 2004, MOD no longer issues service-specific training requirements. Joint doctrine governs all training requirements. The joint task list has been expanded in recent years to connect strategic to operational-level tasks. Having just one joint task list limits confusion.

exercise leads to quicker solutions. The clear delineation of responsibility in MOD means that lessons from the field have a greater chance of being integrated into the training and exercise programs.

Conclusion

The purpose of looking at training activities was not to find the definitive and representative integrated-operations training program but instead (1) to survey what is and what is not being taught now and (2) to assist in the search for models, templates, or examples that could help develop a purpose-built integrated-operations training program.

The data suggest that today we are mostly seeing familiarization training. Courses and programs tend to focus on introducing the basic functions of other governmental organizations rather than a thorough examination of their capabilities and instruction on how DoD personnel might best interact with them. That might indicate that the tasks not being addressed are more difficult or may be more appropriate for the professional integrated operator. This suggests that, for a purpose-built integrated-operations training program, examples exist of how to deal with the easy tasks but that the more difficult tasks will have to be designed from the ground up, rather than tacking on a few integrated-operations elements to the current program in an ad-hoc manner. Chapter Four examines this issue in detail by describing the degree of emphasis that integrated-operations tasks should receive in training activities.

Applying the Methodology

This chapter takes the results of our task-list analysis, discussed earlier, to develop the key elements of an integrated-operations training strategy and outlines some of the challenges to be overcome in implementing it. To focus our research efforts, the team believed that it would be useful to have an operational context in mind. We selected SSTR operations because they require the U.S. military to work with U.S. agencies, allies, and partners in integrated operations.

Aggregate Analysis

Chapters Two and Three discussed how the project team compiled a list of integrated-operations tasks, prioritized the tasks in an SSTR-operations context, assessed the extent to which training contributes to mastering them, and assessed how well the current training regimen addresses them. As the last step of the task analysis, we now assess the priority of tasks within each skill category versus training contribution and coverage for each task or skill category. We did this by overweighting high-priority tasks, high–training contribution tasks, and low training coverage and underweighting low-priority tasks, low–training contribution tasks, and high training coverage within skill categories, then aggregating the results and normalizing to 1.

This approach allows us to simultaneously consider all of the parameters we have considered, which is a necessary step in developing a better way to help DoD prepare to work with non-DoD partners. Combining the priority ranking, the trainability of a task category, and the degree to which the category is covered by the current program, we can offer some recommendations to alter training to better address the need to develop integrated operators. The aggregate results are shown in Figure 4.1.

As a category, conducting operations with and for partners requires the lowest-priority tasks but the highest training coverage. One could speculate that training coverage is high because conducting operations is something at which the military is adept and for which it knows how to train. At the other extreme, providing security cooperation appears to have little training coverage given its priority and susceptibility to a training intervention.

Figure 4.1
SSTR Task Analysis: Aggregate Results

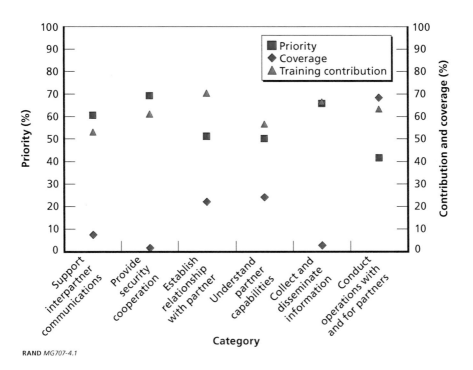

RAND MG707-4.1

Skill-Category Priorities and Strategy Issues

Training can make a contribution to all skill categories, especially to establishing relations with partners. The two skill categories least covered by training, education, and exercise programs are high-priority skill categories. The best-covered category is of low priority but high training contribution. The next two best-covered skill categories are two medium-priority categories.

Several issues emerge from this assessment. One issue is whether all skill categories should be fully covered. In other words, should coverage be driven to "widely taught" for all tasks in every category? If this is desirable, is it feasible, given resource and other constraints? A second related issue is to determine which skill categories should be given emphasis, given that all training coverage cannot reasonably be attained. Should it be the highest-priority skill categories (which have limited coverage)? Where is the greatest benefit to be derived? One might argue that the greatest benefit from new training programs would be in the two highest-priority categories, especially the second highest. For both categories, training can make a significant contribution to task success, and the tasks are not well covered. Providing more training coverage for security-cooperation tasks appears to have the greatest immediate return. A third issue is whether it is desirable to achieve balance across skill categories. If it is good to have each of the skill categories receive equal coverage, DoD would need to reduce coverage for highly covered categories, such as conducting operations with and for partners, and increase coverage for less covered categories, such as providing security cooperation.

Task Priorities and Strategy Issues

Besides making observations based on the skill categories, one can also make judgments at the task level by integrating the three separate assessments. We first selected tasks that were high in priority for SSTR operations, that training could significantly affect, and that had low existing training coverage. (Chapters Two and Three report the assessment of these tasks.) There were 34 of these tasks, as shown in Appendix F. All are from the category of providing security cooperation. These should have the highest priority for new training coverage.[1] In Appendix G are 44 tasks that were high in SSTR-operations priority, high or medium in contribution of training, and low or medium in existing training coverage. These tasks come from all categories, with the exception of conducting operations with and for partners. These should have the second priority for training. Covering these tasks with new training would also bring better balance across the skill categories.

Conversely, at the other extreme are 42 tasks that are low in SSTR-operations priority, that training could not significantly affect, and that had high existing training coverage. These are shown in Appendix H. These tasks come from five of the skill categories, with the exception being the category of conducting operations. These should be considered for having training directed away from them and to the priorities identified here.

Toward a Framework

In this section, we describe some of the factors associated with developing a framework for integrated-operations training. Beginning with a discussion of building capacity, we then develop the elements of a strategy and finally offer alternative frameworks for consideration.

Building Capacity for the Long Term

The previous analysis provided insights on how to prioritize skill groups and tasks to build capability from training. This section raises the issue of capacity. Who is the training audience? How many can or should be trained? How long (either in terms of duration or task repetition) should training be? How enduring will be the benefits of training?

Answers to those questions are governed by choice among competing priorities. Integrating operations is a newly emphasized issue area that, according to some, is a critical priority. But training is constrained by resources, not the least of which are qualified trainers, time, and money. Assuming that military personnel are already fully con-

---

[1]   See later discussion about time and other resources available for training.

sumed by funded training for existing operations and that specialized personnel and units are not created, new training needs for integrated operations will have to take away from other tasks. That is, decisions must be made about how to balance integrated operations with ongoing operational mission needs, such as for major combat operations. If the training audience is officers, noncommissioned officers (NCOs), and petty officers (POs), then other training must be reduced to increase training in integrated operations. Moreover, the training venues examined in earlier chapters do not have sufficient throughput to build integrated-operations knowledge widely in the force. How much throughput (training and education seats balanced with training duration) is needed? Last, having more task repetitions for many types of tasks[2] correlates with better job performance, but increasing training duration for some has the effect of reducing overall throughput. Considerations about depth versus breadth of experience must be addressed.

Elements of an Integrated-Operations Training Strategy
Training is an investment in human-capital formation. Training must integrate with personnel-management practice to be successful. At the heart of human-capital strategy are four steps:

1. *Determine workforce characteristics needed to accomplish integrated operations.* These are some combination of knowledge and skills gained through training, exercises, education, and experience. Typically, the assumption is made that, if training or an assignment is completed, the person has gained the knowledge and skills. A complete determination of specific integrated-operations competencies is useful but not needed. Rather than specifying the knowledge and skill in detail, the proxy for the needed characteristics can be summarized as having had integrated-operations training or integrated-operations experience.
2. *Determine needs (current and future) for these characteristics.* Where (in what positions) are officers, NCOs, or POs with

---

[2] See Glenn et al. (2006, Appendix F) for a discussion of the types of tasks that might rely more on abilities and deep experience.

these characteristics needed? How many of these positions are there? Does this differ by service, by occupation or function, or at different levels of seniority? Beyond the JIACG, are positions for personnel to be added to COCOM or other HQ staffs or to be in operational units? If positions are not added, are these additional duties for personnel already assigned? Do all on a staff or in a unit need the knowledge or skills, or does just a more limited number of specialists need it? Addressing these questions of demand (who, where, and how many) is the critical element before asserting a human-capital strategy to include training and development. It is imperative to know the audience and the size of it to initiate a comprehensive training and development strategy.

3. *Identify officers, NCOs, POs, and enlisted personnel with the needed characteristics.* To our knowledge, no existing database is able to do this. Typically, in the military services, such characteristics are identified through additional skill identifiers (U.S. Army), additional qualification designation (U.S. Navy), skill designator or identifying or reporting member of the service (USMC), and prefixes or other identifiers (USAF). Moreover, in certain cases, specialized databases (e.g., Joint Duty Management Information System) for tracking personnel qualifications are managed by the Joint Staff. Without information on existing integrated-operations capability within the force, it will be difficult to chart an effective strategy.

4. *Use analytical tools to project future availability of personnel with integrated-operations knowledge and skills.* This analysis must include career-management practices, availability of training seats, and availability of qualifying assignments. These are the issues of supply. Moreover, analyses such as this can determine the number of training opportunities (seats) and assignments needed to meet the determined need (demand) for the characteristics. Short- and long-term strategies for eliminating the gaps can then be directed.

These steps in human-capital planning, especially the basic ones of determining demand and supply, are critical as a basis for specifying an overall integrated-operations training and development strategy. Moreover, the training strategy has to be tightly intertwined with personnel-management decisions, because the latter also affect supply of trained and experienced individuals.

Personnel-Management Practice Also Governs Capacity

Personnel-management practices also affect the long-term capacity of the force to increase DoD's capability to work with others. Previous research identified four broad frameworks by which personnel are managed, and each framework has different effects on the distribution of integrated-operations knowledge and experience in the force.

One framework has a *leadership focus* in that personnel who are most likely to advance to senior positions are provided short-duration opportunities to be trained in or experience integrated operations. Because of high demand for these personnel in other billets, they are not likely to have repetitive assignments of this sort. However, because these personnel tend to advance further and stay in careers longer, the senior leadership of the military tends to have at least some experience working with non-DoD partners as a result. The goal of this framework would be to focus limited integrated-operations training resources on those who might most benefit the future military.

In some respects, the opposite of the leadership framework is a *competency* or *cadre framework*. In this framework, personnel are frequently assigned to integrated-operations billets throughout their careers. Personnel gain deep experience but typically do not advance as far or serve as long in their careers as those in the leadership framework. Moreover, because billets are filled repetitively by an experienced cadre, fewer officers have access to such experiences. Overall capacity of the force for integrated operations is lower, but, for those associated with integrated operations, the knowledge and experience are deeper. A benefit of this framework is that training needs might be reduced because further training has less benefit to those with deep experience unless tactics, techniques, and procedures (integrated-operations domain knowledge) change significantly.

A third framework could be labeled a *skill framework*, and its objective is to rotate as many personnel as possible through integrated-operations training and assignments. Broad capacity across the force is the goal of this framework, but it comes at the expense of depth of knowledge and experience. This framework requires widespread access to training opportunities and integrated-operations assignments and may not be feasible for those two reasons, even if advisable for other reasons.

The last framework, which could be labeled the *available-officer framework*, is not so much a desirable framework as the state of the art in personnel management: The available officer is used for integrated operations with or without the necessary training. Rather than having a conscious design for development, the "system" fills a billet with an available person. In the absence of sustained emphasis on the importance of integrated operations, this framework dominates.[3]

All of these frameworks are operating in some degree in the personnel-management system, but the dominance of one or another can affect overall force capacity for integrated operations over the long term.[4]

Overall, resources have to support choices. Beyond resources for training itself, at least two other training-related needs exist. One need is to evaluate training, education, and exercises over longer periods to determine which is most effective for preparing personnel for integrated operations. The training communities in the services evaluate using different processes, but, because working with others outside DoD is not the central feature of most training programs, evaluation might not be consistently conducted or emphasized.

---

[3]   It took 20 years of oversight of joint officer management before Congress was willing to limit the force of law in joint officer management and education in the 2007 authorization act (P.L. 109-364).

[4]   See Thie, Harrell, and Emmerichs (2000) for a fuller discussion of these issues. The 1997 National Defense Panel report (National Defense Panel and DoD, 1997) broached the idea of specialists in IA operations. Congress asked for a report on the feasibility and advisability of establishing a cadre of military officers whose assignments, training, and education would be managed to ensure a viable career track.

The second need is to have a means to track gained integrated-operations knowledge and experience among personnel. The Joint Duty Assignment Management Information System was created to track joint officer education and experience. Should something similar exist for integrated-operations qualifications? Two parts of training transformation, the Joint Knowledge Development and Distribution Capability and the Joint Assessment and Enabling Capability might have insights for how to proceed.

Some balance in training and exercise opportunities across the six skill categories is probably advisable. At this stage of integrated-operations implementation, some knowledge of all appears preferable to deep knowledge in some. This implies more training in providing security cooperation. Specific tasks related to this are in Appendix F.

Whatever program DoD chooses should differentiate between general education and training on the one hand and predeployment training focused on a particular mission on the other. For the former, not training or educating to 100 percent of integrated-operations tasks seems advisable. For the latter, training to all tasks needed for the assignment is advisable.

Time-Phased Approach to Implementation

The first and most critical issue to be resolved is to determine the need for personnel with integrated-operations qualifications. How many? Where? Which services? Which grades? Which occupations or functions? A well-thought-out and -managed data call is needed to determine the scope of need. Tied to this would be a requirement for the services or Joint Staff to identify and manage billets and officers, NCOs, and POs through existing billet and personnel inventory databases.

The second related key issue is to determine career-management practices. Are these skills and knowledge to be managed as additional duties, as separate occupations, and in or out of the mainstream (e.g., unrestricted line or restricted line)? Are these personnel to be comparable to aviators, infantry officers, or surface-warfare officers or more like foreign-area officers? Depending on the nature of those decisions, specification of career practices should be directed by the deputy secretary of defense as they were recently for the Foreign Area Officer Program (DoD, 2005).

Neither of these observations is specifically within the scope of this book, but they are needed precursors to the third issue, which is the training and education strategy. However, resolution of this must await decisions about the prior two. Continuing to emphasize integrated operations in existing training and exercise programs until such time as resolution of the first two issues is clear may be the best alternative. At that time, the number of training and education seats and number of exercises can be determined and resourced.

Last, many state that integrated operations should be a core mission with priority comparable to combat operations. Decisions will need to be made as to which tasks and operations receive less training to devote needed time to train to integrated operations. Once need is determined, training-strategy priority might be to the initiatives that close the largest gaps, that close the most gaps, or that fit the available budget (zero sum). Part of this determination should include the COCOMs' determination of tasks that are critical to their readiness.

Challenges to Conducting Integrated-Operations Training

In addition to surveying what is being done to prepare DoD for integrated operations, we identified a number of obstacles that DoD faces in its efforts to increase its capability to work with others.

Quality

In terms of quality, integrated-operations objectives do not seem to be a priority at most DoD training and exercise events. There is a general lack of available subject-matter experts from other government agencies and countries. IA, multinational, and coalition partner organizations limit their participation in DoD events, which means that DoD participants do not learn the perspectives, strengths, and limitations of their partners. One reason for the difficulty of obtaining partner participation arises from their unwillingness to devote time and money to being little more than "training aids."[5] This presents DoD with something of a dilemma: If the training objectives are set with DoD in mind, it is more difficult to obtain partner participation, but, if the training objectives are altered to meet partner objectives, the event may be of less value to DoD. The challenge, therefore, is to come up with training objectives, scenarios, and event structures that benefit all parties involved in the event. This is easier said than done.

We also observed problems with the way in which training deals with integrated operations. Most courses that focus on integrated-operations topics are devoted to making students more familiar with various parties but do not go into depth about how to work effectively with them. Because there is no requirement for attendance, the training audience is not consistent. For example, such courses as JIMPC target JIACG members but often accept other attendees simply because of their availability. Similarly, while there is a desire to have IA participation, many agencies simply cannot afford to give their limited number of key personnel the time away from the office. In addition, courses that discuss IA cooperation are heavily dependent on guest speakers. While guest speakers often provide a valuable perspective, their qualifications and the quality of their message vary significantly.

Quantity

Our team also observed a number of issues relating to the quantity of integrated-operations preparation. This is an area with endemic shortages of resources relative to the demand placed on it. DoD dwarfs its integrated-operations partners. In terms of organizations, people, and resources, DoD is vast compared to most other agencies in the United States and elsewhere. For example, despite high demand for its involvement, S/CRS can participate in only *one* major DoD exercise each year. Likewise, key allies, such as the UK, Canada, and Australia, operate with relatively limited budgets and cannot afford to participate in as many events as they would prefer, and, for the most part, participa-

---

[5] DoD sometimes turns to civilian contractors, such as retired FSOs, to fill these roles, but these personnel, while experienced, often lack current knowledge of how their former organization works.

tion in many U.S.-sponsored events is beyond the reach of coalition partners from less developed countries. When some of these organizations are able to generate participants for a DoD event, lack of personnel backfill means that their key functions are not performed in their absence, increasing the opportunity costs to their organizations of their further involvement.

In a multitude of DoD events, there are instances of integrated-operations training, exercises, and education, but, in general, IA, multinational, and coalition audiences are underrepresented or not represented at all. Courses are offered infrequently, and class sizes are typically small and proportionally skewed toward the military.

Institutionalization

There are a number of efforts under way to increase DoD effectiveness in integrated operations, but these may be temporary responses to conditions in Iraq, Afghanistan, and elsewhere rather than a deeper commitment that will last beyond the current situation. Certainly, there are efforts throughout DoD to help the U.S. military increase its effectiveness when working with others. For example, the U.S. Army–USMC Counterinsurgency Center at Fort Leavenworth emphasizes the need for cultural awareness and sensitivity to foreign-government partners. The Joint Center for International Security Force Assistance (JCISFA) seeks to gather and promulgate lessons from the field that offer practical advice for how to work with non-DoD partners. These organizations have small budgets and limited staff, and their long-term future is uncertain. There is some coordination between these programs, but this is due largely to the personal relationships of particular officers and is thus not guaranteed to last beyond the next rotational cycle.

The Time Factor

The U.S. military and the USG as a whole face obstacles to increasing their integrated-operations capability. One major obstacle is time. Defense planners are mindful of having limited resources at their disposal, which usually means limited funding. There are opportunity costs for adding new activities to existing programs to create integrated operators. For example, if cultural awareness is added to a soldier's predeployment training, one of two things must happen. Potentially, the instruction on cultural awareness will displace some other instruction on a different activity. It may be the case that the displaced activity is of only minor importance, but, as more and more integrated-operations skills are added to the curriculum, trade-offs will need to be made. An option, then, is to simply make the course longer. This, however, also results in an opportunity cost. The time factor, then, is a constraint on the ability to develop activities that build integrated-operator skills.

Cultural Differences

We mentioned previously that other resource disparities between the United States and its partners limit opportunities for interaction and that there is tension between satisfying DoD and partner needs. In addition, cultural differences make it difficult for DoD to work with others. Some cultural differences are more pronounced in IA cooperation than in international cooperation. For example, the DoD emphasis on meticulous and disciplined planning is rarely matched by another organization. To DoD, a plan is a multifaceted, detailed effort that considers a wide variety of environmental conditions, assumptions, courses of action, and outcomes. To other USG agencies, a plan is often a one-page description of goals.

Organization

DoD's organizational setup also contributes to the problem of preparing for integrated operations and creates a dilemma for IA organizations when it comes to where to expend their scarce resources. Title 10 gives the services responsibility, inter alia, to train and educate military forces. However, since it is the responsibility of the COCOMs to conduct actual operations, it often makes better sense for IA organizations to work with them instead. Thus, the services, which produce the trainers and educators, are somewhat removed from the partners with whom soldiers, sailors, airmen, and marines need to work. To address this disconnect, the draft joint training manual calls for COCOMs to designate "high interest training issues" to influence service training policy (Sprenger, 2007).

Cooperation

One final obstacle is common to the phenomenon of cooperation. Different organizations have different goals and interests. While this obstacle is most evident in operations with international partners, different perspectives also complicate IA cooperation. This is not a new problem, nor is it easy to overcome. It will inherently place limits on the quality and quantity of DoD efforts to prepare itself to work more effectively with others.

CHAPTER FIVE

Recommendations

To increase DoD's ability to work with others, it needs an integrated personnel, leadership, and training strategy. For the most part, the team found that there are "pockets" of integrated-operations training being conducted at various levels. This is not surprising given the as-yet-uncoalesced thinking on this topic. Still, the activities that do exist could form the basis for an integrated-operations training program.

DoD will need to make some difficult choices to build an effective program. The obvious comparison with joint operations suggests that a similar emphasis on integrated operations may help to effectively surmount such challenges. An integrated solution to increasing integrated-operations capability would need to go beyond training to include officer and enlisted recruitment, selection, and career management.

Determine the Nature and Scope of Demand for Integrated Operations

This book provides a view of the landscape with respect to integrated-operations training. This is merely a first step, but a necessary one, toward the creation of a training program designed to increase integrated-operations capability. To progress further, DoD would have to know what sort of integrated-operations capability it needs. It would need a demand analysis to determine how many officers or NCOs from which services, grades, and functions it needs to increase integrated-operations capabilities. There are likely to be tiers of integrated-operations knowledge that the joint force requires: a general baseline of knowledge that a great many officers possess and a smaller group of more-specialized officers. DoD needs a demand analysis to determine the scope of each of these tiers and how they should be composed. Once this analysis is complete, Office of the Under Secretary of Defense (OUSD) for Personnel and Readiness (P&R) should be able to construct a plan and develop a time-phased approach to implement it.

Formalize an Integrated-Operations Task List

In the absence of such a demand analysis, there are some steps that OUSD P&R should take. These steps could be taken sequentially. First, it should review the integrated-operations task list in Appendix A as well as the integrated-operations categories and publish an integrated-operations task list. Publication of the list would encourage discussion among COCOMs, the services, the Joint Staff, and others in the training community and would help raise awareness of the need to better prepare DoD to work with others.

Emphasize Training for the Highest-Priority Tasks

Second, OUSD P&R should review the task lists in Appendixes F through H, send them to training organizations, and require them to include the tasks in future events. Appendixes F and G list tasks that should receive additional attention. In the event that additional resources are *not* provided for integrated-operations training, Appendix H lists tasks that could receive less emphasis using criteria described in Chapter Four. If the level of resources is increased, tasks in Appendixes F and G could receive more emphasis without having to decrease emphasis on the Appendix H tasks.

Maintain and Increase Visibility of Integrated-Operations Training

Third, OUSD P&R should consider maintaining a database of ongoing training, exercises, and professional military-education courses at the service and joint levels that focus on integrated operations. The Joint Training Information Management System database could serve as a tool to this end, but, as it is presently used, units rarely report training objectives, which renders it ineffective and makes it difficult, if not impossible, to accurately assess the state of the training program.

OUSD P&R might also develop qualitative and quantitative measures of effectiveness (MOEs) to begin to analyze the utility of these tasks and their application over time to ensure adequate funding for the most effective programs. Such MOEs could be included in a database to which trainers and program managers have access. Moreover, OSD should seek to gain additional insights into the training activities undertaken by key allies, such as the UK and others, on how they build their own capacity for integrated operations, as some of these practices might be transferable to the U.S. military.

Provide Stable Funding Sources for Innovative Programs

Fourth, once OUSD P&R has determined the most successful approaches, it should advocate for stable funding for innovative programs. We have discussed several possible candidates in Chapter Three. Some of these programs are the result of the labor of a few visionary officers and other leaders and lack dedicated funding streams. After conducting further analysis on the effectiveness of each integrated-operations training program, OUSD P&R should select a few of these programs and ensure that they receive the support necessary to continue their work. Without such action, these programs are likely to wither with time as the defense community focuses on new issues.

In addition to expanding programs that work, we recommend that OUSD P&R examine the utility of new training methods. As we have noted, one of the difficulties inherent in integrated operations is the lack of available partners with whom to train. The use of distributed simulation tools could help mitigate this problem by making it easier to collaborate more frequently with non-DoD partners and by simulating partners when none is available or when it is inappropriate to include them.

Integrated Operations in Perspective

Ultimately, DoD's ability to work with others is a function of more than just preparatory efforts, such as training, exercises, and education. In addition to these steps, DoD would need to look at officer and enlisted recruitment and career management, and even areas outside the personnel sphere, such as doctrine and organization. The task is immense. Instead of waiting for others to step forward, however, the training community should do what it can to help prepare the joint force to work with others. Doing so would reap considerable gains in operational effectiveness, which would, in turn, contribute to the furtherance of U.S. national-security interests.

Integrated-Operations Task List

Table A.1 describes the tasks relevant to integrated operations.

Table A.1

Integrated-Operations Task List

Task Number[a]	Task	Description	CAT[b]	PRI[c]	CON[d]
ST 8.2.2	Coordinate civil affairs in theater	To coordinate those activities that foster relationships between theater military forces and civil authorities and people in a friendly country or area.	1	H	C
ST 8.5.3	Establish theater IA-cooperation structure	To establish formal and informal relationships with other USG departments and agencies in the theater for the mutual exchange of information and support.	1	H	M
ART 7.10.3	Maintain community relations	To assist civil-affairs personnel in conducting (planning, preparing, executing, and assessing) community-relations programs as resources permit.	1	H	C
ST 8.5.3.2	Support regional IA activities	To conduct direct liaison with various agencies or departments for coordination, preparation, and implementation of regional IA activities.	1	M	C
OP 5.7.1	Ascertain national or agency agenda	To bring out the unstated agendas of each participant in a joint or multinational effort; to understand each nation or agency's individual goals within the framework of a joint or multinational effort.	1	M	M
SN 5.6	Provide PA worldwide	To advise and assist the secretary of defense, CJCS, and combined chiefs in an alliance in telling the military's story to both internal and external audiences.	1	M	C
SN 8.1.9	Cooperate with and support NGOs	To work with and arrange for a mutually beneficial relationship between DoD and NGOs.	1	M	L

Table A.1—Continued

Task Number[a]	Task	Description	CAT[b]	PRI[c]	CON[d]
OP 4.4.6	Provide religious-ministry support in the JOA	To coordinate the provision of religious support among components of a JTF. Additional activities include assisting NGOs with humanitarian-assistance (HA) programs.	1	L	L
ART 6.14.3	Identify local resources, facilities, and support	To identify, locate, and assist in the acquisition of local resources, civilian labor, facilities, and other support that tactical organizations require to accomplish their missions.	1	M	C
SN 4.2.9	Acquire HN support	To negotiate and contract for support and services from an HN for U.S. forces in a theater and within the United States if in response to homeland-security missions.	1	M	M
ST 8.3.1	Arrange stationing for U.S. forces	To obtain approval for and to house and dispose forces to best support peacetime presence and military operations.	1	H	M
SN 3.1.3	Support establishment of access and storage agreements	To support the combatant commander's efforts to obtain agreements for periodic access by U.S. personnel and units and for the permanent stationing ashore or afloat of selected equipment and supplies.	1	H	C
ST 8.1.2	Promote regional security and interoperability	To work with allies within the framework of military alliances to improve or secure U.S. position within the region.	1	M	C
SN 3.1.2	Coordinate periodic and rotational deployments, port visits, and military contacts	To collaborate with other U.S. departments and agencies and the U.S. Congress and to work with foreign governments to allow for U.S. combat, support, and training units; individual service members; and DoD civilians to visit foreign nations.	1	L	C

Table A.1—Continued

Task Number[a]	Task	Description	CAT[b]	PRI[c]	COND[d]
ST 8.1.1	Enhance regional politico-military relations	To strengthen and promote alliances through support of regional relationships.	1	L	C
SN 8.1.3	Support peace operations	To support peace operations through national-level coordination of the three general areas: diplomatic action, traditional peacekeeping, and forceful military action.	2	H	L
ST 8.2.5	Coordinate nation-assistance support	To support and assist in development of nations, normally in conjunction with DOS, an ally, or both.	2	H	C
ST 8.2.6	Coordinate military civic-action assistance	To coordinate with or assist HN forces on projects useful to the local population.	2	M	M
ST 8.2.7	Assist in restoration of order	To halt violence and reinstitute peace and order.	2	H	M
ST 8.2.9	Coordinate theater FID activities	To coordinate the participation of civilian and military agencies of a government in any of the action programs taken by another government to free and protect its society from subversion, lawlessness, and insurgency.	2	H	L
OP 4.4.5	Train joint forces and personnel	To train replacements and units, especially newly rebuilt units, in the theater of operations. In MOOTW, this activity includes training assistance for friendly nations and groups.	2	H	C
OP 4.7.2	Conduct civil-military operations (CMO) in the JOA	To conduct activities that foster the relationship of the military forces with civilian authorities and population and that develop favorable emotions, attitudes, or behavior among neutral, friendly, or hostile groups.	2	H	M

Table A.1—Continued

Task Number[a]	Task	Description	CAT[b]	PRI[c]	CON[d]
OP 4.7.4	Transition to civil administration	To implement the transition from military administration in a region to UN or civil administration in the region.	2	H	M
OP 4.7.7	Conduct FID	To provide assistance in the operational area to friendly nations facing threats to their internal security.	2	H	C
ART 6.14.6	Establish temporary civil administration (friendly, allied, and occupied enemy territory)	To establish a temporary civil administration (at the direction of the national command authority [NCA]) until existing political, economic, and social conditions stabilize in enemy territory or in friendly territory where there is a weak or ineffective civil government.	2	H	M
SN 8.1.8	Provide support to FID in theater	To work with U.S. agencies and foreign governments to provide programs to support action programs to free and protect the foreign nation's society from subversion, lawlessness, and insurgency.	2	H	M
ST 8.2.3	Coordinate foreign humanitarian assistance	To anticipate and respond to national, multinational, and IA requests for assistance for events that occur outside the United States and its territories and possessions.	2	M	C
ART 6.14.6.2	Provide public-administration support	To provide liaison to the military forces. Survey and analyze the operation of local governmental agencies: their structure, centers of influence, and effectiveness.	2	H	C
ART 6.14.6.4	Provide public-education support	To provide technical advice and help in planning and implementing needed education programs.	2	M	L
OP 5.7.7	Conduct civil-administration operations	To conduct, when approved by the secretary of defense, certain functions of civil government.	2	H	L

Table A.1—Continued

Task Number[a]	Task	Description	CAT[b]	PRI[c]	COND[d]
ART 6.14.6.12	Provide public legal support	To establish supervision over local judiciary system, establish civil administration courts, and help in preparing or enacting necessary laws for the enforcement of U.S. policy and international law.	2	H	C
ART 8.3.2.1	Provide indirect support to FID	Indirect support emphasizes the principles of HN self-sufficiency and builds strong national infrastructures through economic and military capabilities.	2	H	C
ART 8.3.2.2	Provide direct support to FID (not involving combat operations)	Direct support (not involving combat operations) involves the use of U.S. forces providing direct assistance to the HN civilian populace or military.	2	H	C
NTA 4.8	Conduct civil-affairs activities in area	To conduct those activities that embrace the relationship of the military forces with civilian authorities and population in a friendly country or area or in an occupied country or area when military forces are present.	2	H	M
ART 6.14.6.3	Provide public electronic-communication support	To manage communication resources, public and private, including postal services, telephone, telegraph, radio, television, and public warning systems.	2	H	C
SN 9.3	Conduct arms-control support activities	To implement intrusive arms-control inspections to fulfill treaty obligations, including conducting on-site inspections.	2	L	M
SN 9.4	Support weapons of mass destruction (WMD) nonproliferation and counterproliferation activities and programs	To implement and coordinate WMD nonproliferation and counterproliferation activities that respond to U.S. policy and strategy objectives for combating WMD proliferation and WMD terrorism.	2	L	M

Table A.1—Continued

Task Number[a]	Task	Description	CAT[b]	PRI[c]	CON[d]
SN 8.1.2	Support-nation assistance	To support and assist in developing other nations, normally in conjunction with DOS or a multinational force or both and, ideally, through the use of HN resources.	2	H	C
SN 8.1.4	Support military-civic action	To support the use of predominantly indigenous military forces on projects useful to the local populace in fields contributing to economic and social development.	2	M	C
ART 6.14.6.6	Provide public-health support	To estimate needs for additional medical support required by the civilian sector.	2	M	C
ART 6.14.6.7	Provide public-safety support	To coordinate public-safety activities for the military force.	2	M	C
ART 6.13	Conduct internment and resettlement activities	Includes activities performed by units when they are responsible for interning enemy prisoners of war (EPWs) and civilian detainees.	2	H	M
SN 8.1.6	Provide civil-affairs support policy	To provide policy on activities that embrace the relationship between a nation's military forces and its civil authorities and people in a friendly country or area or occupied country or area, when military forces are present.	2	H	C
ST 8.2.1	Coordinate security-assistance activities	To provide defense articles, military training, and advisory assistance and other defense-related services.	2	H	M
ART 6.14.6.5	Provide public-finance support	To provide technical advice and assistance regarding budgetary systems, monetary and fiscal policies, revenue-producing systems, and treasury operations.	2	M	L

Table A.1—Continued

Task Number[a]	Task	Description	CAT[b]	PRI[c]	CON[d]
ART 6.14.6.11	Provide civil-defense support	To ensure that an adequate civil-defense structure exists. Advise, assist, or supervise local civil-defense officials.	2	H	C
SN 3.1.4	Coordinate joint and multinational training events	To coordinate, schedule, and conduct designated joint and multinational training events.	2	M	M
SN 8.1.1	Provide security assistance	To provide defense articles, military training, and other defense-related services by grant, credit, or cash sales to further national policies and objectives.	2	H	M
SN 8.2.2	Support other government agencies	To support non-DOD agencies. Support includes military support to civil authorities and civilian law-enforcement agencies (LEAs), counterdrug operations, combating terrorism, noncombatant evacuation, and building a science and technology base.	2	H	C
ST 7.1.7	Establish joint METL (JMETL)	To analyze applicable tasks derived through mission analysis of joint operation plans and external directives and select for training only those tasks that are essential to accomplish the organization's wartime mission.	2	L	M
ST 8.2.4	Coordinate humanitarian- and civic-assistance programs	To assist nations in the theater with medical, dental, and veterinary care and construction and repair of basic transportation, wells, sanitation, and other public facilities.	2	M	M
OP 4.7.1	Provide security assistance in the JOA	To provide friendly nations or groups with defense articles, military training, and other defense-related services by grant, loan, credit, or cash sales in furtherance of national policies and objectives within the JOA.	2	H	C

Table A.1—Continued

Task Number[a]	Task	Description	CAT[b]	PRI[c]	CON[d]
ART 6.1.10	Provide supplies for civilian use (class X)	To provide material to support nonmilitary programs, such as agriculture and economic development.	2	M	M
ART 7.7.2.2.4	Provide customs support	To perform tactical actions that enforce restrictions on controlled substances and other contraband violations that enter or exit an area of operation (AO).	2	L	M
ART 8.3.3	Conduct security assistance	Security assistance refers to a group of programs that support U.S. national policies and objectives by providing defense articles, military training, and other defense-related services to foreign nations by grant, loan, credit, or cash sales.	2	H	M
ART 8.3.6.6	Provide training support to counterdrug efforts	Training support to LEAs and HNs includes basic military skills, such as basic marksmanship, patrolling, mission planning, and medical and survival skills.	2	M	C
ART 8.3.6.8	Provide research, development, and acquisition support to counterdrug efforts	The Army Counterdrug Research, Development, and Acquisition Office makes military research, development, and acquisition efforts available to LEAs.	2	L	C
ART 8.4.3.2.1	Provide support to domestic preparedness (for weapons of mass destruction)	The National Domestic Preparedness Office, under the Federal Emergency Management Agency (FEMA), orchestrates the national domestic-preparedness effort.	2	L	M
ART 8.4.3.3.3	Provide general support to civil law enforcement	Provide limited military support to LEAs. DoD may direct Army forces to provide training to federal, state, and local civilian LEAs.	2	L	C

Table A.1—Continued

Task Number[a]	Task	Description	CAT[b]	PRI[c]	CON[d]
SN 8.3	Coordinate military activities within the IA process	To work with representatives of the other executive departments and agencies to resolve issues involving both overseas and domestic operations.	3	H	M
ST 8.2.10	Coordinate multinational operations within theater	To coordinate with allies, coalition partners, and appropriate international organizations to ensure mutual support and consistent effort in the theater.	3	M	L
ST 8.5.3.4	Coordinate planning for IA activities	Integrate military operations with organizations representing other agencies.	3	H	L
OP 4.7.5	Coordinate politico-military support	To coordinate and support politico-military activities among military commands, DoD and other USG agencies, and friendly governments and groups within the JOA.	3	H	M
OP 5.7.4	Coordinate plans with non-DoD organizations	To facilitate exchange of operational information, ensure coordination of operations among coalition or agency forces, and provide a forum in which routine issues can be resolved informally among staff officers.	3	M	L
OP 6.5.5	Integrate HN security forces and means	To integrate and synchronize HN police, fire departments, military internal-security forces, communication infrastructure, constabulary, rescue agencies, and penal institutions into the security plan for the operational area.	3	H	L
NTA 4.8.3	Provide IA coordination	To coordinate all civil affairs with the appropriate U.S. agencies and follow their direction as appropriate.	3	M	M

Table A.1—Continued

Task Number[a]	Task	Description	CAT[b]	PRI[c]	CON[d]
MCTL 5.4.3	Coordinate and integrate joint, multinational, and IA support	To coordinate with elements of the joint force, allies and coalition partners, and other government agencies to ensure cooperation and mutual support, a consistent effort, and a mutual understanding of the joint force commander's priorities.	3	H	C
SN 8.1.10	Coordinate actions to combat terrorism	To coordinate action to preclude, preempt, and resolve terrorist actions across the threat spectrum, including antiterrorism and counterterrorism.	3	H	L
SN 8.3.1	Coordinate and control policy for the conduct of operations	To work with the other partners in the IA process to ensure that all ideas going forward to the president have been fully understood by all IA participants.	3	H	C
OP 5.4.5	Coordinate and integrate component, theater, and other support	To coordinate with allies and coalition partners and U.S. component commands.	3	M	M
OP 5.6.1	Integrate operational information operations (IO)	To integrate the offensive and defensive actions involving information, information-based processes, information systems, and psychological-operations (PSYOP) activities.	3	M	M
OP 5.7.2	Determine national and agency capabilities and limitations	To take action to determine multinational-force or agency capabilities, strengths, and weaknesses to match missions with capabilities and exploit special or unique capabilities of member forces or agencies.	3	H	C
ART 3.3.2.4	Conduct PSYOP	To integrate planned psychological messages, products, and actions into combat operations or in support of MOOTW.	3	H	C

Table A.1—Continued

Task Number[a]	Task	Description	CAT[b]	PRI[c]	CON[d]
OP 4.7.6	Coordinate civil affairs in the JOA	To coordinate those activities that foster relationships of operational forces with local civil authorities and people in a friendly country or area.	3	H	M
SN 7.4.4	Conduct joint, multinational, interoperability, and IA training of assigned forces	To plan, execute, and analyze joint, multinational, interoperability, and IA training for assigned forces to perform those tasks and capabilities to specified conditions and standards in support of the commander's requirements.	3	M	C
ST 8.5.3.1	Establish JIACG to facilitate IA activities	To establish and operate a theater JIACG to plan, coordinate, and assist the unified commander in execution of joint IA operations.	3	M	L
OP 5.7.6	Coordinate coalition support	To coordinate coalition-support activities to provide the combined-force commander the means to acquire coalition-force status and capabilities.	3	M	M
NTA 4.8.4	Coordinate with NGOs	To coordinate civil affairs with appropriate NGOs, including private voluntary organizations (PVOs).	3	M	L
SN 7.4.6	Provide joint, multinational, interoperability, and IA training for other than assigned forces	To assist with analysis, planning, and execution of joint, multinational, interoperability, and IA training for other than assigned forces.	3	M	L
ST 8.4.2	Combat terrorism	To produce effective anticipatory and offensive measures to defeat transnational terrorist organizations; prevent WMD acquisition, development, or use by terrorist organizations; and develop partner countries' capacity to detect and defeat terrorists.	3	H	M

Table A.1—Continued

Task Number[a]	Task	Description	CAT[b]	PRI[c]	CON[d]
ST 8.5.1	Coordinate and integrate policy for the conduct of theater operations	To work within the country team and other forums to provide support to the programs of other USG departments and agencies within the theater.	3	H	C
ST 9.1	Integrate efforts to counter weapon and technology proliferation in theater	To integrate support of DoD and other government agencies to prevent, limit, or minimize the introduction of chemical, biological, radiological, nuclear, and explosive (CBRNE) weapons; new, advanced weapons; and advanced weapon technologies to a region.	3	L	C
ST 9.2	Coordinate counterforce operations in theater	To positively identify and select CBRNE-weapon targets, such as acquisition, weaponization, facility preparation, production, infrastructure, exportation, deployment, and delivery systems.	3	L	M
ST 8.2.11	Coordinate with and support NGOs in theater	To work with and arrange for a mutually beneficial relationship between the combatant commander and NGOs operating within the theater.	3	M	M
ST 8.2.12	Coordinate with and support PVOs in theater	To work with and arrange for a mutually beneficial relationship between the combatant commander and PVOs operating within the theater.	3	M	M
ST 8.2.13	Coordinate countermine activities	To coordinate U.S. forces support for countermine activities in the theater with NGOs, HNs, and USG agencies.	3	M	C
MCTL 4.6.3	Plan, coordinate, and monitor EPWs, civilian internees, and U.S. military prisoner operations	To plan, coordinate and monitor the collection, processing, and transfer of EPWs, civilian internees, and U.S. military prisoners.	3	H	C

Table A.1—Continued

Task Number[a]	Task	Description	CAT[b]	PRI[c]	CON[d]
SN 7.4.1	Coordinate JMETL or agency METL (AMETL) development	To provide methodology and policy for establishing combatant-commander JMETL and combat-support AMETL.	3	L	C
ST 8.3.2	Establish bilateral or multilateral arrangements	To establish, in anticipation of requirements to conduct operations with friends and allies outside an alliance command structure, mutually agreed procedures.	3	M	M
ST 8.4.1	Advise and support counterdrug operations in theater	To support counterdrug operations through the establishment of theater JTFs or elements of multijurisdictional forces in support of LEAs and HN forces.	3	M	M
OP 5.5.3	Integrate joint-staff augmentees	To integrate augmentees into existing staff structure to form a joint staff to support USJFCOM.	3	L	C
OP 5.7.3	Develop multinational intelligence and information-sharing structure	To optimize each member nation's intelligence and information capabilities, determine what information may be shared with multinational partners, and provide member forces a common intelligence picture tailored to their requirements.	3	M	M
OP 5.7.5	Coordinate HN support	To coordinate HN support in the JOA to ensure the most effective fit with military and contracted support capabilities.	3	M	L
ART 6.3.1	Provide movement control	To plan, route, schedule, control, coordinate, and provide in-transit visibility of personnel, units, equipment, and supplies moving via all modes of transportation (less pipeline) over air and surface lines of communication.	3	M	C
ART 6.14.6.9	Provide public-welfare support	To determine the type and amount of welfare supplies needed for emergency relief.	3	M	C

Table A.1—Continued

Task Number[a]	Task	Description	CAT[b]	PRI[c]	CON[d]
ART 6.14.6.13	Provide civilian supply support	To determine the availability of local supplies for civil and military use.	3	M	C
ART 6.14	Provide economic and commerce support	To determine the availability of local resources for military and civilian use.	3	M	C
ART 6.14.6.16	Provide property-control support	To identify private and public property and facilities available for military use and recommend policy and procedures to obtain them.	3	M	C
NTA 2.1.7	Supervise intelligence, counterintelligence (CI), and reconnaissance operations	To monitor and assess the effectiveness of intelligence, CI, and reconnaissance operations.	3	M	C
MCTL 4.4.6	Coordinate combat and civil-military engineering plans and operations	To review operation plans and combat- and civil-engineer support plans and approve USMC-force engineer plans.	3	M	C
ST 8.5.3.3	Assess military participation during IA activities	To plan, coordinate, and implement assessment methodology to determine effectiveness of military participation during IA activities and share information with participants.	3	M	L
OP 7.5	Integrate JOA ISR with CBRNE situation	To integrate the CBRNE-weapon situation into command, control, communication, computer, and ISR (C4ISR) systems in the JOA.	3	L	C
ART 6.14.6.8	Provide public-transportation support	To identify the modes and capabilities of transportation systems available in the civilian sector.	3	M	M

Table A.1—Continued

Task Number[a]	Task	Description	CAT[b]	PRI[c]	CON[d]
ART 8.3.6.4	Provide planning support to counterdrug efforts	U.S. Army personnel support counterdrug planning of both LEAs and HNs.	3	M	M
ART 8.3.2.3	Conduct combat operations in support of FID	Combat operations include offensive and defensive operations conducted by U.S. forces to support an HN fight against insurgents or terrorists.	4	H	M
NTA 6.3.3	Combat terrorism	To perform defensive and offensive measures to reduce vulnerability of individuals and property to terrorist acts; to prevent, deter, and respond to terrorism.	4	H	C
ART 8.3.1.3	Conduct operations in support of diplomatic efforts	U.S. Army forces support diplomatic efforts to establish peace and order before, during, and after conflicts.	4	H	C
ART 7.7.2.2.2	Conduct criminal investigations	To investigate offenses against U.S. forces or property committed by persons subject to military law.	4	M	C
NTA 4.8.1	Support peace operations	To provide logistical, medical, and other services to mixed populations in support of disaster relief, HA, and civil-action programs.	4	M	M
ART 6.13.2	Conduct populace and resource control	To provide security for a populace, denying personnel and materiel to the enemy, mobilize population and material resources, and detect and reduce the effectiveness of enemy agents.	4	H	C
ART 8.3.7.2	Conduct antiterrorism activities	Antiterrorism reduces the vulnerability of individuals and property to terrorist attacks, including limited response and containment by local military forces.	4	H	C

Table A.1—Continued

Task Number[a]	Task	Description	CAT[b]	PRI[c]	CON[d]
MCTL 6.5.3	Combat terrorism	To perform defensive and offensive measures to reduce vulnerability of individuals and property to terrorist acts.	4	H	C
TA 6.2	Execute personnel-recovery operations	To execute personnel-recovery operations using component, joint, multinational, and multiagency personnel-recovery capabilities to report, locate, support, recover, and debrief and reintegrate U.S. military and other designated personnel.	4	L	L
ST 8.3.4	Obtain multinational support against nonmilitary threats	To identify and obtain cooperation and support of allies and friends for protection against nonmilitary threats to civilian and military personnel and to key facilities in the theater.	4	M	C
OP 6.1.2	Integrate joint and multinational operational aerospace defense	To implement an integrated air-defense system from all available joint and multinational operational defense forces (aircraft, missiles, air-defense artillery).	4	L	L
ART 5.3.7	Conduct defensive IO	To plan, coordinate, and integrate policies and procedures, operations, personnel, and technology to protect and defend information and information systems.	4	L	C
OP 5.5.4	Deploy joint-force HQ advance element	To deploy elements of the HQ into the operational area in advance of the remainder of the joint force.	4	L	M
ART 6.14.5	Resettle refugees and displaced civilians	To estimate the number of dislocated civilians, their points of origin, and anticipated direction of movement.	4	H	M
ART 7.7.2.2.5	Provide refugee and displaced-civilian movement control	To assist, direct, or deny the movement of civilians whose location, direction of movement, or actions may hinder operations.	4	H	M

Table A.1—Continued

Task Number[a]	Task	Description	CAT[b]	PRI[c]	CON[d]
OP 6.2.9	Coordinate personnel recovery	To integrate joint, multinational, and IA capabilities and coordinate PR operations to report, locate, support, recover, and reintegrate.	4	L	L
TA 6.4	Conduct noncombatant evacuation	Tactical operations involving land, sea, and air forces to evacuate U.S. dependents, USG employees, and private citizens (U.S. and third-country) from locations in a foreign country or HN to a designated area within the theater.	4	L	L
ART 8.3.5.1	Provide operational, logistic, and training support to insurgencies	To provide support to insurgencies in the form of equipment, training, and services.	4	L	M
OP 6.2	Provide protection for operational forces, means, and noncombatants	To safeguard friendly centers of gravity and operational force potential by reducing or avoiding the effects of enemy operational-level (tactical risks) actions.	4	H	M
TA 7.1	Conduct mission operations in a CBRNE environment	To apply principles of avoid, protect, and decontaminate to joint forces operating in proximity to the threat or actual use of CBRNE. Includes coordination and standardization of warning and reporting between joint and multinational forces.	4	M	C
SN 8.1.11	Support countermine activities	To support the elimination of the threat to noncombatants and friendly military forces from mines, booby traps, and other explosive devices.	4	M	C
OP 7.4	Coordinate consequence management (CM) in the JOA	To coordinate support for IA essential services and activities required to manage and mitigate damage resulting from the employment of CBRNE weapons or release of toxic industrial materials (TIMs) or contaminants.	4	L	M

Table A.1—Continued

Task Number[a]	Task	Description	CAT[b]	PRI[c]	CON[d]
ART 5.3.8.4	Perform counter–human intelligence (HUMINT)	Counter-HUMINT is designed to defeat or degrade threats to HUMINT-collection capabilities.	4	M	C
SN 1.2.2	Provide forces and mobility assets	To provide the transportation assets (e.g., road, rail, sealift, and airlift) required in an operational configuration for the movement of forces and cargo.	4	H	M
ST 8.3.3	Arrange sustainment support for theater forces	To obtain sustainment support from sources other than the U.S. military. This activity includes HN support, logistic civil augmentation, third-country support, and captured materiel.	4	M	M
OP 4.7.3	Provide support to DoD and other government agencies	To provide support to DoD, Joint Staff, other services, Defense Information Systems Agency, Defense Logistics Agency, Defense Threat Reduction Agency, DOS, USAID, U.S. Information Agency, civil governments, and other related agencies.	4	H	C
ART 6.14.6.10	Provide public-works and facility support	To coordinate public-works and utility support for military operations.	4	M	M
ART 8.3.4	Conduct humanitarian and civic assistance	Humanitarian- and civic-assistance programs consist of assistance provided in conjunction with military operations and exercises.	4	M	M
ART 8.4.3.1.2	Provide humanitarian relief	Humanitarian relief focuses on lifesaving measures that alleviate the immediate needs of a population in crisis.	4	M	M
NTA 4.7	Perform civil military-engineering support	To repair and construct facilities and lines of communication and to provide water, utilities, and other related infrastructure.	4	M	M
NTA 6.3.2.3	Manage refugees and refugee camps	To collect, process, evaluate, safeguard, house, and release refugees.	4	M	C

Table A.1—Continued

Task Number[a]	Task	Description	CAT[b]	PRI[c]	CON[d]
MCTL 6.4.1	Coordinate management of refugees	To collect, process, evaluate, safeguard, house, and release refugees.	4	M	C
MCTL 6.4.2	Conduct military law enforcement	To enforce military law and order and collect and evacuate EPWs.	4	H	C
ST 6.2.6.4	Establish and coordinate theaterwide CI requirements	To establish and coordinate activities or actions to provide protection against espionage, sabotage, or assassinations conducted by or on behalf of foreign governments or international terrorist activities in the theater.	4	M	L
OP 6.2.6	Conduct evacuation of noncombatants from the JOA	To use JOA military and HN resources for the evacuation of U.S. military dependents, USG civilian employees, and private citizens (U.S. and third-country nationals).	4	L	M
ART 2.1.1	Conduct mobilization of tactical units	Mobilization is the process by which U.S. Army tactical forces or part of them are brought to a state of readiness for war or other national emergency.	4	L	C
ART 2.1.1.1	Conduct alert and recall	Units and individuals receive mobilization or alert orders, individuals assigned to the unit are notified of the situation, and all individuals report to the designated location at the designated time with designated personal items.	4	L	C
ART 2.1.1.2	Conduct home-station mobilization activities	This task involves the activities of reserve-component units at home station after receiving a mobilization order followed by entry onto federal active duty or other C2 changes.	4	L	C

Table A.1—Continued

Task Number[a]	Task	Description	CAT[b]	PRI[c]	CON[d]
ART 5.3.6.1	Provide protective services for selected individuals	To protect designated high-risk individuals from assassination, kidnapping, injury, or embarrassment.	4	H	C
ART 6.14.6.15	Provide food and agriculture support	To provide advice and assistance in establishing and managing crop-improvement programs, agricultural training, use of fertilizers and irrigation, livestock improvement, and food processing, storage, and marketing.	4	L	M
ART 7.7.2.2	Provide law and order	To ensure lawful and orderly environment and suppress criminal behavior.	4	H	M
ART 8.3.6.2	Support HN counterdrug efforts	U.S. Army forces support counterdrug efforts indirectly through civilian agencies of the USG and the civilian or military organizations of the host country.	4	M	M
ART 8.3.6.5	Provide logistic support to counterdrug efforts	U.S. Army forces can assist LEAs or HNs during their conduct of counterdrug operations with logistic management and execution.	4	M	C
ART 8.3.6.7	Provide personnel support to counterdrug efforts	U.S. Army forces may provide a variety of individuals or units to support HN and IA counterdrug efforts.	4	M	C
ART 8.3.8	Perform noncombatant-evacuation operations (NEOs)	NEOs relocate threatened civilian noncombatants from locations in a foreign nation to secure areas.	4	L	M
ART 8.3.10	Conduct a show of force	Shows of force are flexible deterrence options designed to demonstrate U.S. resolve.	4	L	M

Table A.1—Continued

Task Number[a]	Task	Description	CAT[b]	PRI[c]	CON[d]
ART 8.4.3.1.1	Provide disaster relief	Disaster relief restores or recreates essential infrastructure. It includes establishing and maintaining the minimum safe working conditions, less security measures, necessary to protect relief workers and the affected population.	4	M	L
NTA 1.4.8.1	Conduct alien-migrant interdiction operations	To intercept alien migrants at sea, rescue them from unsafe conditions, and prevent their passage to U.S. waters and territory.	4	L	M
NTA 4.4.2.4	Provide billeting to noncombatant evacuees	To use available military resources to provide accommodations, food, and emergency supplies to U.S. dependents, USG civilian employees, and private citizens (U.S. and third-nation) who have been evacuated from the AO.	4	L	C
MCTL 1.2.0.18	Conduct NEOs	Operations directed by DOS, DoD, or other appropriate authority, whereby noncombatants are evacuated from foreign countries when their lives are endangered by war, civil unrest, or natural disaster to safe havens or to the United States.	4	L	C
SN 8.1.5	Conduct foreign humanitarian and civic assistance	To conduct assistance to relieve or reduce the results of natural or human-caused disasters, including CM, or other endemic conditions that might present a serious threat to life or that can result in great damage to or loss of property.	4	M	C
OP 4.6.4	Provide law enforcement and prisoner control	To collect, process, evacuate, and intern EPWs and to enforce military law and order in the communication zone (COMMZ) and in support of operational-level commander's campaigns and major operations.	4	H	C
ART 5.3.5.4	Conduct area-security operations	Area security is a security operation conducted to protect friendly forces, installations, routes, and actions within a specific area.	4	M	M

Table A.1—Continued

Task Number[a]	Task	Description	CAT[b]	PRI[c]	CON[d]
ART 5.3.5.5	Conduct local security operations	To take measures to protect friendly forces from attack, surprise, observation, detection, interference, espionage, terrorism, and sabotage.	4	H	M
ART 5.3.5.5.5	Conduct critical installations and facility security	To secure and protect installations and facilities from hostile action.	4	M	M
ART 5.3.6.2	React to a terrorist incident	To implement measures to treat casualties, minimize property damage, restore operations, and expedite the criminal investigation and collection of lessons learned from a terrorist incident.	4	M	L
ART 5.3.6.3	Reduce vulnerabilities to terrorist acts and attacks	To reduce personnel vulnerability to terrorism by understanding the nature of terrorism, knowing current threats, identifying vulnerabilities to terrorist acts, and implementing protective measures against terrorist acts and attack.	4	M	M
ART 7.7.2.2.1	Perform law enforcement	To assist commanders through conducting law-enforcement operations, including the maintenance of liaison activities and support of the training of other DoD police organizations and HN police authorities.	4	H	M
NTA 6.3.2	Conduct military law-enforcement support (afloat and ashore)	To enforce military law and order and collect, evacuate, and intern EPWs.	4	H	C
NTA 6.5.1	Provide disaster relief	To deliver disaster relief, including personnel and supplies, and provide a mobile, flexible, rapidly responsive medical capability for acute medical and surgical care.	4	M	C

Table A.1—Continued

Task Number[a]	Task	Description	CAT[b]	PRI[c]	CON[d]
SN 8.2.3	Support evacuation of noncombatants from theaters	To provide for the use of military and civil, including HN, resources for the evacuation of U.S. dependents, USG civilian employees, and private citizens (U.S. and third-nation).	4	L	C
SN 8.2.4	Assist civil defense	To assist other federal agencies and state governments in mobilizing, organizing, and directing the civil population to minimize the effects of enemy action or natural and technological disasters on all aspects of civil life.	4	L	M
SN 9.2.2	Coordinate CM	To contain, mitigate, and repair damage resulting from the intentional use or accidental release of a CBRNE weapon or a TIM.	4	L	C
SN 9.2.4.1	Support CM	To provide subject-matter expertise on site or through technical reachback to support CM, accident and incident response, and mitigation for domestic and foreign CM incidents and events.	4	L	C
ST 8.4.3	Coordinate evacuation and repatriation of noncombatants from theater	To use all available means to evacuate U.S. dependents, USG civilian employees, and private citizens (U.S. and third-country) from the theater and support the repatriation of appropriate personnel to the United States.	4	L	C
ST 8.4.5	Coordinate civil support in the United States	To plan for and respond to domestic requests for assistance from other USG and state agencies in the event of civil emergencies, such as natural and human-caused disasters, CM, civil disturbances, and federal work stoppages.	4	L	C
ST 9.5	Coordinate CM in theater	To coordinate support for planning and conducting CM in theater. Task includes establishing liaison with necessary government agencies, regional NGOs, international organizations, and regional military commands that contribute resources to CM operations.	4	M	M

Table A.1—Continued

Task Number[a]	Task	Description	CAT[b]	PRI[c]	CON[d]
OP 4.7.8	Establish disaster-control measures	To take measures before, during, or after hostile action or natural disasters to reduce probability of damage, minimize its effects, and initiate recovery.	4	M	M
ART 5.3.5.4.1	Conduct rear-area and base security operations	Rear-area and base security operations are a specialized area-security operation.	4	M	M
ART 5.3.8.5	Perform counter-signal intelligence (SIGINT)	Counter-SIGINT is designed to defeat or degrade threats to SIGINT-collection capabilities.	4	M	C
ART 5.3.8.6	Perform countermeasurement and signature	Countermeasurement and signature intelligence is designed to defeat or degrade threat measurement and SIGINT-collection capabilities.	4	M	C
NTA 6.2.1	Evacuate noncombatants from area	To use available military and civilian resources (including HN resources) to evacuate U.S. dependents, USG civilian employees, and private citizens (U.S. and third-nation) from the AO.	4	L	M
ART 8.4.3.2.2	Protect critical assets	Hostile forces may attack facilities essential to society, the government, and the military.	4	H	C
ART 8.4.3.2.3	Respond to WMD incidents	Other government agencies have primary responsibility for responding to domestic terrorist and WMD incidents.	4	L	M
ART 8.4.3.3.1	Support U.S. Department of Justice counterterrorism activities	When directed by the NCA, U.S. Army forces may provide assistance to the Department of Justice in the areas of transportation, equipment, training, and personnel.	4	L	C

Table A.1—Continued

Task Number[a]	Task	Description	CAT[b]	PRI[c]	CON[d]
ART 8.4.3.3.2	Conduct civil-disturbance operations	The U.S. Army assists civil authorities in restoring law and order when state and local LEAs are unable to control civil disturbances.	4	L	M
ART 8.4.3.4	Provide community assistance	Community assistance is a broad range of activities that provide support and maintain a strong connection between the military and civilian communities.	4	L	C
OP 2.1.1	Determine and prioritize operational PIRs	To assist USJFCOM in determining and prioritizing its PIRs. In MOOTW, it includes helping and training HNs to determine their IRs, such as in COIN operations.	5	H	M
ART 7.2.6	Communicate with non-English speaking forces and agencies	To communicate orally, nonverbally, in writing, or electronically in the appropriate language of allied, HN, NGO, and indigenous forces and agencies to accomplish all C2 requirements.	5	H	C
SN 8.3.3	Establish IA cooperation structures	To work within the IA process, ensuring that knowledgeable personnel represent the views of the Joint Chiefs of Staff and the combatant commanders.	5	H	L
OP 3.2.2.1	Employ PSYOP in the JOA	To plan and execute operations to convey selected information and indicators to foreign audiences in the AO to influence their emotions, motives, or objective reasoning.	5	H	M
ST 5.5.1	Plan and integrate theaterwide IO	To plan theaterwide IO, integrating military operations and non-DoD USG activities.	5	M	L
OP 2.1.2	Determine and prioritize operational IRs	To identify those items of information that must be collected and processed to develop the intelligence required by the commander's PIR.	5	H	C

Table A.1—Continued

Task Number[a]	Task	Description	CAT[b]	PRI[c]	CON[d]
ART 6.14.6.18	Provide cultural-affairs support	To provide information to military forces on the social, cultural, religious, and ethnic characteristics of the local populace.	5	H	M
OP 5.1.5	Monitor strategic situation	To be aware of and understand national and multinational objectives, policies, goals, other elements of national and multinational power, political aim, and the GCC's strategic concept and intent.	5	H	C
OP 5.8.1	Manage media relations in the JOA	To provide support to the commander in ensuring the timely and correct telling of the command's story.	5	H	L
OP 5.8.3	Conduct community-relations programs in the JOA	Within the JOA, conduct community-relations programs in coordination with civil affairs that support direct communication with local, national, and international publics, as applicable.	5	H	L
ART 7.10	Conduct PA operations	To advise and assist the commander and command (or HN, in MOOTW) in PA planning.	5	H	M
ART 7.10.1	Execute information strategies	To identify affected internal and external audiences and their information requirements.	5	H	C
NTA 5.8	Provide PA services	To advise and assist the commander, associated commands, and coalition partners in providing information to internal and external audiences by originating print and broadcast news material and assisting with community-relations projects.	5	H	M
NTA 6.1.2	Conduct perception management	To convey or deny selected information and indicators to foreign audiences to influence their emotions, motives, and objective reasoning.	5	H	C

Table A.1—Continued

Task Number[a]	Task	Description	CAT[b]	PRI[c]	CON[d]
TA 5.6	Employ tactical IO (TIO)	TIO employed by joint services produce tactical information and gain, exploit, defend, or attack information or information systems.	5	L	C
OP 6.2.12	Provide counter-PSYOP	To conduct activities to identify adversary psychological-warfare operations contributing to situational awareness and serve to expose adversary attempts to influence friendly populations and military forces.	5	H	M
ART 6.14.2	Locate and identify population centers	To locate and identify population centers in the area of interest and anticipate population movements that may occur in response to future combat operations.	5	M	C
ART 5.3.8	Conduct tactical CI in the AO	CI is designed to defeat or degrade threat intelligence–collection capabilities.	5	M	C
ART 5.3.8.2	Perform CI	To gather information and conduct activities to protect against espionage, other intelligence activities, sabotage, or assassinations conducted by or on behalf of foreign governments or elements thereof.	5	M	C
ART 6.14.6.17	Provide civil print information support	To advise, assist, supervise, control, or operate civil print information agencies.	5	M	M
ART 5.3.7.2	Conduct counterpropaganda	To establish plans and procedures to counter enemy PSYOP based on an effective PA and education program to expose, discount, and inform targeted audiences of threat-propaganda initiatives.	5	H	C
ART 8.3.9	Conduct arms-control operations	U.S. Army forces normally conduct arms-control operations to support arms-control treaties and enforcement agencies.	5	L	C

Table A.1—Continued

Task Number[a]	Task	Description	CAT[b]	PRI[c]	CON[d]
NTA 1.4.8.2	Conduct maritime counterdrug operations	To coordinate with all applicable agencies to detect and monitor vessel and air traffic and provide vessels and qualified boarding teams to intercept, board, inspect, search, and, as appropriate, seize vessels suspected of smuggling drugs.	5	M	C
SN 2.1.5	Determine national strategic-intelligence issues	To identify issues involving intelligence collection, planning, exploitation, production, and dissemination that requires resolution by the secretary of defense, director of central intelligence, or military-intelligence boards.	5	M	M
ST 8.1.4	Develop multinational intelligence and information-sharing structure	To enhance each member nation's intelligence and information capabilities through development of sharing structure.	5	M	M
OP 2.1.3	Prepare operational collection plan	To develop a collection plan that will satisfy the commander's intelligence and CI requirements.	5	H	L
OP 2.2	Collect and share information	To gather information from operational and tactical sources on operational and tactical threat forces and their decisive points (and related high-payoff targets, such as CBRNE weapon production, infrastructure, and delivery systems).	5	H	C
ART 7.4.2.2	Enhance friendly decisionmaking	To leverage information management that supports making more precise and timely decisions than the enemy.	5	H	C
NTA 6.3.1.6	Conduct surveillance-detection operations	To identify, locate, and help counter the enemy's intelligence, espionage, sabotage, subversion, and terrorist-related activities, capabilities, and intentions to deny the enemy the opportunity to take action against friendly forces.	5	H	C

Table A.1—Continued

Task Number[a]	Task	Description	CAT[b]	PRI[c]	CON[d]
MCTL 2.1.8	Provide tactical CI and HUMINT support	To sufficiently suppress or defeat the enemy's intelligence-collection, terrorism, and sabotage efforts to allow the regiment to conduct its mission with the element of surprise and with minimal losses.	5	H	M
MCTL 2.2	Collect information	To gather combat and intelligence data to satisfy the identified requirements.	5	H	M
SN 2.2.1	Collect information on strategic situation worldwide	To obtain information and data from all sources on the strategic situation.	5	M	M
SN 3.4.7.1	Produce counterterrorism intelligence	CI input to threat assessments includes the current or projected capability of a foreign intelligence service.	5	M	L
ST 5.6	Develop and provide PA in theater	To develop and provide to the combatant commander and allied partners a program for telling the theater and combined command's story to audiences both internal and external.	5	H	C
OP 2.1.4	Allocate intelligence resources in the JOA	To assign adequate resources to theater and JTF intelligence organizations to permit the accomplishment of assigned intelligence tasks. Includes requesting support and the reallocation of additional assets from allied countries.	5	M	C
OP 5.1.3	Determine commander's critical information requirements	To determine the critical information that a commander requires to understand the flow of operations and to make timely and informed decisions.	5	H	M

Table A.1—Continued

Task Number[a]	Task	Description	CAT[b]	PRI[c]	CON[d]
OP 6.2.11	Provide counterdeception operations	To neutralize, diminish the effects of, or gain advantage from a foreign deception operation.	5	L	M
ART 1.1.2	Perform situation development	Situation development is a process for analyzing information and producing current intelligence about the enemy and environment during operations.	5	H	M
ART 1.1.3	Provide intelligence support to force protection	To provide intelligence in support of protecting the tactical forces' fighting potential so that it can be applied at the appropriate time and place.	5	H	C
ART 1.1.4.1	Collect police information	Collection of police information involves all available collection capabilities.	5	M	C
ART 1.2.3	Conduct area studies of foreign countries	Study and understand the cultural, social, political, religious, and moral beliefs and attitudes of allied, HN, or indigenous forces to assist in accomplishing goals and objectives.	5	H	M
ART 1.4.2.1.1	Provide intelligence support to PSYOP	This task identifies the cultural, social, economic, and political environment of the AO.	5	H	C
ART 1.4.2.3.1	Provide intelligence support to CMO	This task allows military intelligence organizations to collect and provide information and intelligence products concerning foreign cultural, social, economic, and political elements within an AO in support of CMO.	5	H	C
ART 1.4.2.3.2	Provide intelligence support to PA	This task identifies the coalition and foreign public physical and social environment, as well as world, HN national, and HN local public opinion.	5	H	M

Table A.1—Continued

Task Number[a]	Task	Description	CAT[b]	PRI[c]	COND[d]
ART 3.3.2.4.1	Develop PSYOP products	To develop products to support offensive, defensive, stability, and support operations.	5	H	C
ART 5.3.5.6.1	Identify essential elements of friendly information (EEFI)	To identify friendly vulnerabilities that are exploitable by enemies and potential adversaries. Include recommendations concerning countermeasures and corrective action.	5	M	M
ART 6.14.4	Advise commanders of obligations to civilian population	To develop, in conjunction with the staff judge advocate, requirements and guidance for military personnel concerning the treatment of the civilian population.	5	M	M
ART 6.14.6.1	Provide arts, monuments, and archive support	To prepare a list and map overlay showing the location of significant cultural properties requiring special protection.	5	M	C
ART 7.2.1.1	Collect friendly force information requirements	To collect data about friendly forces from the information environment for processing, displaying, storing, and disseminating to support C2 functions.	5	M	C
MCTL 2.1.1	Conduct area or country studies	To obtain information on the social environment, on the political environment, and the economic environment.	5	H	C
MCTL 2.1.6	Plan and coordinate SIGINT support	To plan and coordinate SIGINT support from national, theater, JTF, and other component assets.	5	M	C
MCTL 2.1.9	Plan and coordinate operational-level CI	To plan and coordinate CI and HUMINT policy, doctrine, and procedures.	5	M	C
MCTL 2.2.2.1	Conduct aerial reconnaissance and surveillance	Air reconnaissance provides information for the formulation of plans and policies at the national and international level.	5	M	C

Table A.1—Continued

Task Number[a]	Task	Description	CAT[b]	PRI[c]	CON[d]
SN 2.8	Provide CI support	To provide CI support to CJCS, COCOMs, services, and other agencies.	5	M	C
SN 8.1.7	Coordinate information-sharing arrangements	To arrange for the selected release and disclosure of unclassified and classified information in support of multinational operations and exercises.	5	H	M
SN 8.3.2	Conduct information management in the IA process	To ensure that the maximum information is made available to all participants in the IA process.	5	M	L
SN 8.3.5	Coordinate DoD and government IO	To work with the services, COCOMs, and civil and military agencies on issues involving offensive and defensive IO.	5	M	C
ST 8.5.2	Facilitate U.S. information exchange in the region	To ensure the free flow of information among USG departments and agencies in the theater.	5	M	C
ST 9.6	Integrate theater ISR with the CBRNE-weapon situation	To interface the CBRNE-weapon situation with theater C4ISR systems. Includes the processing of information from U.S. and multinational strategic, operational, and tactical sources.	5	L	L
ART 1.1.4.2	Conduct the police-information assessment process (PIAP)	PIAP is a tool used to contribute to police intelligence operations (PIO).	5	M	C
ART 1.1.4.3	Develop police-intelligence products	PIO use the intelligence cycle to produce actionable police-intelligence products used by military police leaders in tactical and nontactical environments.	5	M	M

Table A.1—Continued

Task Number[a]	Task	Description	CAT[b]	PRI[c]	CON[d]
ART 3.3.2.4.2	Produce PSYOP products	To prepare PSYOP products for distribution and dissemination to a target audience.	5	M	C
ART 7.10.2	Facilitate media operations	To provide assistance to media that are covering operations.	5	M	C
MCTL 5.2.8	Coordinate PA services	To advise and assist the commander, associated commands, and coalition partners in providing information to internal and external audiences by originating and assisting civilian news media.	5	H	C
ART 8.3.6.1	Support detection and monitoring of drug shipments	To provide aerial and ground reconnaissance to support counterdrug operations by LEAs.	5	M	M
ST 8.1.3	Develop HQ or organizations for coalitions	To establish, as appropriate, headquarters, organizations, or both to support operations in war or MOOTW.	6	H	M
ST 8.2.8.2	Establish and coordinate a peacekeeping infrastructure	To establish, preserve, and maintain peace through an infrastructure of military or civilian personnel (or both).	6	M	C
ART 6.14.1	Provide interface or liaison between U.S. military forces and local authorities and NGOs	To facilitate CMO by providing interface between U.S. military forces and HN, foreign authorities, foreign military forces, or NGOs.	6	H	M

Table A.1—Continued

Task Number[a]	Task	Description	CAT[b]	PRI[c]	CON[d]
MCTL 5.2.6	Provide liaison to local authorities and NGOs	To facilitate CMO by providing interface between U.S. military forces and HN, foreign authorities, or NGOs.	6	H	C
OP 5.6.3	Control IO	To monitor and adjust the operational IO efforts during execution.	6	M	M
OP 5.5.6	Establish or participate in task forces	To establish or participate in a functional or single-service task force established to achieve a specific limited objective. This task force may be single-service, joint, or multinational.	6	L	M
OP 5.5.1	Develop a joint-force C2 structure	To establish a structure for C2 of subordinate forces.	6	H	C
OP 5.5.2	Develop a joint-force liaison structure	To establish a structure to maintain contact or intercommunication between elements of the joint force to ensure mutual understanding and unity of purpose and action.	6	M	L
ART 8.3.6.3	Provide command, control, communication, computer, and intelligence support to counterdrug efforts	U.S. Army personnel and equipment may assist LEAs and HNs in designing, implementing, and integrating command, control, communication, computer, and intelligence systems.	6	M	M

[a] ST = strategic theater. ART = article in the AUTL. OP = operational. SN = strategic national. NTA = article in the UNTL. TA = tactical.

[b] Skill category as shown in Table 2.2 in Chapter Two.

[c] Task priority. Assessed as high (H), medium (M), or low (L), as discussed in Chapter Two.

[d] Training contribution. Assessed as critical (C), moderate (M), or little (L), as discussed in Chapter Two.

Analysis of Training Objectives: Headquarters Staff

Table B.1 describes training objectives for HQ staff.

Table B.1
Analysis of Training Objectives: Headquarters Staff

Category	Task Number[a]	Task	ITEA	JSOU		
				SOFIACC	Terrorism Response	JIMPC
Establish relationship with partner	ST 8.2.2	Coordinate civil affairs in theater				
	ST 8.5.3	Establish theater IA-cooperation structure	USG capabilities and coordination exercise	The NSC and the IA process	IA roles	The IA process
			Coordination case studies and lessons learned			JIACG
	ST 8.5.3.2	Support regional IA activities	USG capabilities and coordination exercise	The NSC and IA process	IA roles	The IA process
	OP 5.7.1	Ascertain national or agency agenda	IA coordination overview: NSC and HSC	The NSC and IA process	IA roles	The IA process
	SN 8.1.9	Cooperate with and support NGOs				NGOs and transnational corporations

Table B.1—Continued

| Category | Task Number[a] | Task Description | ITEA | JSOU | | | |
				SOFIACC	Terrorism Response	JIMPC
Establish relationship with partner (continued)	ST 8.1.2	Promote regional security and interoperability				
	SN 3.1.2	Coordinate periodic and rotational deployments, port visits, and military contacts				
	ST 8.1.1	Enhance regional politico-military relations				
	ST 8.2.6	Coordinate military civic-action assistance				
Provide security cooperation	ST 8.2.9	Coordinate theater FID activities		Shaping the environment: security assistance and FID		

Table B.1—Continued

Category	Task Number[a]	Task Description	ITEA	JSOU		
				SOFIACC	Terrorism Response	JIMPC
Provide security cooperation (continued	OP 4.4.5	Train joint forces and personnel				
	OP 4.7.2	Conduct CMO in the JOA				
	OP 4.7.7	Conduct FID		Shaping the environment: security assistance and FID		
	SN 8.1.8	Provide support to FID in theater		Shaping the environment: security assistance and FID		
	ART 8.3.2.1	Provide indirect support to FID		Shaping the environment: security assistance and FID		

Table B.1—Continued

Category	Task Number[a]	Task Description	ITEA	JSOU SOFIACC	JSOU Terrorism Response	JIMPC
Provide security cooperation (continued)	ART 8.3.2.2	Provide direct support to FID (not involving combat operations)		Shaping the environment: security assistance and FID		
	NTA 4.8	Conduct civil-affairs activities in area				
	SN 8.1.4	Support military-civic action				
	SN 8.1.6	Provide civil-affairs support policy				
	ST 8.2.1	Coordinate security-assistance activities		Shaping the environment: security assistance and FID		

Table B.1—Continued

Category	Task Number[a]	Task Description	ITEA	JSOU		
				SOFIACC	Terrorism Response	JIMPC
Provide security cooperation (continued)	SN 3.1.4	Coordinate joint and multinational training events				
	SN 8.1.1	Provide security assistance		Shaping the environment: security assistance and FID		
	SN 8.2.2	Support other government agencies	USG capabilities and coordination exercise	Collaboration with intelligence agencies	Roles of other USG agencies	Intelligence support to the IA effort
			Coordination with state and local governments	The embassy country team		The country team
			Coordination-challenge case studies and lessons learned	JIACG and JIATF		JIACG

Table B.1—Continued

Category	Task Number[a]	Task Description	JSOU			
			ITEA	SOFIACC	Terrorism Response	JIMPC
Provide security cooperation (continued)	OP 4.7.1	Provide security assistance in the JOA				
	ART 8.3.3	Conduct security assistance				
Understand partner capabilities	SN 8.3	Coordinate military activities within the IA process	IA coordination overview: the NSC and HSC

Coordination with state and local governments

USG capabilities and coordination exercise

Coordination-challenge case studies and lessons learned | Collaboration with other agencies

JIACG and JIATF

The embassy country team | | The JIACG concept

Ambassador or country team and military |

Table B.1—Continued

Category	Task Number[a]	Task Description	ITEA	JSOU		
				SOFIACC	Terrorism Response	JIMPC
Understand partner capabilities (continued)	ST 8.2.10	Coordinate multinational operations within theater				IA exercise (Coherent Kluge)
	ST 8.5.3.4	Coordinate planning for IA activities	IA coordination overview: the NSC and HSC USG capabilities and coordination exercise Coordination-challenges case studies and lessons learned	Collaboration with other agencies SOF-IA collaboration exercise		

Table B.1—Continued

Category	Task Number[a]	Task Description	ITEA	SOFIACC	Terrorism Response	JIMPC
				JSOU		
Understand partner capabilities (continued)	OP 4.7.5	Coordinate politico-military support	USG capabilities and coordination exercise	Collaboration with other agencies		IA exercise (Coherent Kluge)
			Coordination-challenges case studies and lessons learned (U.S. only)	SOF-IA collaboration exercise		
	OP 5.7.4	Coordinate plans with non-DoD organizations		Collaboration with other agencies		IA exercise (Coherent Kluge)
				SOF-IA collaboration exercise		
	NTA 4.8.3	Provide IA coordination		Collaboration with other agencies		IA exercise (Coherent Kluge)
				SOF-IA collaboration exercise		

Table B.1—Continued

Category	Task Number[a]	Task Description	ITEA	SOFIACC	Terrorism Response	JIMPC
				JSOU		
Understand partner capabilities (continued)	MCTL 5.4.3	Coordinate and integrate joint, multinational, and IA support	USG capabilities and coordination exercise	Collaboration with other agencies		IA exercise (Coherent Kluge) (U.S. only)
			Coordination-challenge case studies and lessons learned (U.S. only)	SOF-IA collaboration exercise (U.S. only)		IA exercise (Coherent Kluge)
	SN 8.3.1	Coordinate and control policy for the conduct of operations	USG capabilities and coordination exercise			
			Coordination-challenge case studies and lessons learned			
	OP 5.4.5	Coordinate and integrate component, theater, and other support				

Table B.1—Continued

Category	Task Number[a]	Task Description	JSOU			
			ITEA	SOFIACC	Terrorism Response	JIMPC
Understand partner capabilities (continued)	OP 5.7.2	Determine national and agency capabilities and limitations	USG capabilities and coordination exercise Coordination-challenge case studies and lessons learned (U.S. only)	The IA process (U.S. only)		The IA process (U.S. only)
	OP 4.7.6	Coordinate civil affairs in the JOA				
	SN 7.4.4	Conduct joint, multinational, interoperability, and IA training of assigned forces				
	ST 8.5.3.1	Establish JIACG to facilitate IA activities	Coordination-challenges case studies and lessons learned	JIACG and JIATF		The JIACG concept

Table B.1—Continued

Category	Task Number[a]	Task Description	ITEA	JSOU			JIMPC
				SOFIACC	Terrorism Response		
Understand partner capabilities (continued)	NTA 4.8.4	Coordinate with NGOs					NGOs and transnational corporations
	SN 7.4.6	Provide joint, multinational, interoperability, and IA training for other than assigned forces					
	ST 8.4.2	Combat terrorism			Terrorism-response case studies		
	ST 8.2.11	Coordinate with and support NGOs in theater					NGOs and transnational corporations
	ST 8.5.3.3	Assess military participation during IA activities	Coordination-challenge case studies and lessons learned	Collaboration with other agencies JIACG and JIATF The embassy country team			IA players in complex contingencies

Table B.1—Continued

Category	Task Number[a]	Task Description	ITEA	SOFIACC	JSOU Terrorism Response	JIMPC
Conduct operations with and for partners	ART 8.3.2.3	Conduct combat operations in support of FID			Security assistance and FID	
	NTA 6.3.3	Combat terrorism			Terrorism-response case studies	
	ART 8.3.7.2	Conduct antiterrorism activities			Terrorism-response case studies	
	MCTL 6.5.3	Combat terrorism			Terrorism-response case studies	
	ART 8.3.5.1	Provide operational, logistic, and training support to insurgencies			Security assistance and FID	

Table B.1—Continued

Category	Task Number[a]	Task Description	ITEA	JSOU		
				SOFIACC	Terrorism Response	JIMPC
Conduct operations with and for partners (continued)	OP 4.7.3	Provide support to DoD and other government agencies	USG capabilities and coordination exercise Coordination-challenges case studies and lessons learned	Collaboration with other agencies SOF-IA collaboration exercise		IA exercise (Coherent Kluge)
Collect and disseminate information	SN 8.3.3	Establish IA cooperation structures	USG capabilities and coordination exercise Coordination-challenge case studies and lessons learned	The embassy country team JIACG and JIATF Collaboration with other agencies SOF-IA collaboration exercise		Ambassador, country team, and military The JIACG concept IA exercise (Coherent Kluge)

Table B.1—Continued

Category	Task Number[a]	Task Description	ITEA	JSOU		
				SOFIACC	Terrorism Response	JIMPC
Collect and disseminate information (continued)	ART 6.14.6.18	Provide cultural-affairs support				
	OP 2.2	Collect and share information		Collaboration with intelligence agencies		Intelligence support to the IA effort
	MCTL 2.1.1	Conduct area and country studies				
Support interpartner communications	MCTL 5.2.6	Provide liaison to local authorities and NGOs				NGOs and transnational corporations

NOTE: Green = widely taught. Yellow = narrowly taught. Red = not taught.

[a] ST = strategic theater. OP = operational. SN = strategic national.

Analysis of Training Objectives: JTF Training

Table C.1 describes objectives for JTF training.

Table C.1
Analysis of Training Objectives: JTF Training

Category	Task Number[a]	UE 05-2 Operation Enduring Freedom (OEF) 6 MRX	UE 05-3 Multi National Corps–Iraq MRX	Combined Joint Task Force–Horn of Africa 07-1	Sharp Focus 07
Establish relationship with partners	OP 1.2.4		Construct mechanisms to engage the local populace		Enhance working relationship with potential coalition partners
	OP 3.3	UE 05-2 OEF 6 MRX for Southern European Task Force (SETAF) predeployment for OEF, January 2005			
	OP 4.7		Integrate CMO into IO across strategic, operational, and tactical lines	Provide politico-military support to other nations, groups, and government agencies	
	OP 4.7.4		Integrate CMO into IO across strategic, operational, and tactical lines		
	OP 4.7.6			Coordinate civil affairs in the JOA	

Table C.1—Continued

Category	Task Number[a]	UE 05-2 Operation Enduring Freedom (OEF) 6 MRX	UE 05-3 Multi National Corps–Iraq MRX	Combined Joint Task Force–Horn of Africa 07-1	Sharp Focus 07
Establish relationship with partners (continued)	OP 5.7.5	Establish a civil-military operation center (CMOC) in the combined joint operation area (CJOA), and conduct CMOs			
	OP 5.7	Execute transition plan to pass responsibility to civilian authorities	Coordinate transition between coalition, HN, and government officials		
			Facilitate development of good civilian governance, including local or municipal and regional governmental and judicial systems		

Table C.1—Continued

Category	Task Number[a]	UE 05-2 Operation Enduring Freedom (OEF) 6 MRX	UE 05-3 Multi National Corps–Iraq MRX	Combined Joint Task Force–Horn of Africa 07-1	Sharp Focus 07
Establish relationship with partners (continued)	OP 4.7.3	Demonstrate ability to coordinate multiple HA-activity requirements among IO, NGOs, HN, and CJTF capabilities and forces			
	OP 5.7.4	Provide support to DoD and other government agencies			
		Coordinate plans with non-DoD organizations			
Provide security cooperation	OP 1.2.4		Assess local support requirements and plan, resource, and direct reconstruction efforts to provide an environment conducive to CJTF		

Table C.1—Continued

Category	Task Number[a]	UE 05-2 Operation Enduring Freedom (OEF) 6 MRX	UE 05-3 Multi National Corps–Iraq MRX	Combined Joint Task Force–Horn of Africa 07-1	Sharp Focus 07
Provide security cooperation (continued)	OP 4.7.2		Coordinate transfer of U.S. materiel to HN, IO, or NGOs		
			Provide logistical support to emergency humanitarian operations		
	OP 5.4.5	Integrate coalition partners and HN contracting capacity into engineering			
	OP 5.7			Coordinate and integrate joint and multinational and IA support	
Understand partner capabilities	OP 3.2.2	Develop PSYOP concept to support operation			
	OP 3.2.2.1	Coordinate PSYOP employment			

Table C.1—Continued

Category	Task Number[a]	UE 05-2 Operation Enduring Freedom (OEF) 6 MRX	UE 05-3 Multi National Corps–Iraq MRX	Combined Joint Task Force–Horn of Africa 07-1	Sharp Focus 07
Understand partner capabilities (continued)	OP 4.7.2	Establish initial liaison and coordination with coalition nations	Establish a CMOC in the CJOA and CMOs		Train SETAF to conduct peace operations in accordance with (IAW) contingency plans 4242 and 4265
		Identify HN capabilities pertinent to CMOs	Identify HN capabilities pertinent to CMOs		
Conduct operations with and for partners	OP 1.2.4.8			Conduct unconventional warfare in the JOA	
	OP 3.2.2.1	Coordinate PSYOP employment			
	OP 6.2			Provide protection for operational forces, means, and noncombatants	

Table C.1—Continued

Category	Task Number[a]	UE 05-2 Operation Enduring Freedom (OEF) 6 MRX	UE 05-3 Multi National Corps–Iraq MRX	Combined Joint Task Force–Horn of Africa 07-1	Sharp Focus 07
Collect and disseminate information	OP 2.1			Direct operational intelligence activities	
	OP 2.1.4	Theater and JTF intelligence resources are identified and allocated appropriately within JOA			
	OP 2.2			Collect and share operational information	
	OP 2.5			Disseminate and integrate operational intelligence sustainment	
	OP 5.2	Respond to a negative media report and provide analysis of situation			

Table C.1—Continued

Category	Task Number[a]	UE 05-2 Operation Enduring Freedom (OEF) 6 MRX	UE 05-3 Multi National Corps–Iraq MRX	Combined Joint Task Force–Horn of Africa 07-1	Sharp Focus 07
Collect and disseminate information (continued)	OP 5.6	Plan, synchronize, and coordinate IO with relevant parties	Plan, synchronize, and coordinate IO with CJTF HQ and across CJTF staff, IA, and coalition partners	Coordinate operational IO	
	OP 5.7	Establish communication with multinational forces	Establish communication with IA and multinational forces		
		Integrate other USG agencies in the PA planning process	Establish and implement foreign-disclosure procedures for U.S. members of CJTF		
	OP 5.7.6		Establish liaison with coalition forces		

Table C.1—Continued

Category	Task Number[a]	UE 05-2 Operation Enduring Freedom (OEF) 6 MRX	UE 05-3 Multi National Corps–Iraq MRX	Combined Joint Task Force–Horn of Africa 07-1	Sharp Focus 07
Support interpartner communications	OP 5.5			Establish, organize, and operate a joint-force HQ	Train a JTF HQ IAW U.S. European Command (USEUCOM) directive 55-11
					Train SETAF as the core of a CJTF HQ
					Train USEUCOM and service components in combined C2 and mutual support procedures

NOTE: Green = key focus. Yellow = somewhat. Red = not at all.

[a] OP = operational.

Analysis of Training Objectives: MSTP and BCTP

Table D.1 describes training objectives for MSTP and BCTP.

Table D.1
Analysis of Training Objectives: MSTP and BCTP

Category	II MEF	II MEF	I MEF Forward	III MEF	II MEF Forward	101 ABN DIV	10 MTN DIV	III Corps	82 ABN DIV	25 ID(10)	38 ID(11)
Establish relationship with partners											
Provide security cooperation											
Understand partner capabilities											
Conduct operations with and for partners											
Collect and disseminate information											
Support interpartner communication											

NOTE: Green = key focus. Yellow = somewhat. Red = not at all. MEF = marine expeditionary force. ABN DIV = airborne division. MTN DIV = mountain division.

Analysis of Training Objectives: Military Advisers

Table E.1 describes training objectives for military advisers.

Table E.1
Analysis of Training Objectives: Military Advisers

Category	Task Number[a]	Task	6SOS	MiTT Training
Establish relationship with partner	ST 8.2.2	Coordinate civil affairs in theater	CMO	
	ST 8.5.3	Establish theater IA-cooperation structure		
	ST 8.5.3.2	Support regional IA activities		
	OP 5.7.1	Ascertain national or agency agenda	Adviser techniques	
	SN 8.1.9	Cooperate with and support NGOs		
	ST 8.1.2	Promote regional security and interoperability	Adviser techniques; Political and cultural integration techniques; Regional orientation	Cultural awareness
	SN 3.1.2	Coordinate periodic and rotational deployments, port visits, and military contacts	Adviser techniques	Cultural awareness

Table E.1—Continued

Category	Task Number[a]	Task	6SOS	MiTT Training
Establish relationship with partner (continued)	ST 8.1.1	Enhance regional politico-military relations	Cross-cultural communication	Cultural awareness
Provide security cooperation	ST 8.2.6	Coordinate military civic-action assistance	CMO	
	ST 8.2.9	Coordinate theater FID activities		
	OP 4.4.5	Train joint forces and personnel		
	OP 4.7.2	Conduct CMO in the JOA	CMO	
	OP 4.7.7	Conduct FID		
	SN 8.1.8	Provide support to FID in theater		
	ART 8.3.2.1	Provide indirect support to FID		
	ART 8.3.2.2	Provide direct support to FID (not involving combat operations)		
	NTA 4.8	Conduct civil-affairs activities in area	CMO	

Table E.1—Continued

Category	Task Number[a]	Task	6SOS	MiTT Training
Provide security cooperation (continued)	SN 8.1.4	Support military-civic action	CMO	
	SN 8.1.6	Provide civil-affairs support policy	CMO	
	ST 8.2.1	Coordinate security-assistance activities	Security-assistance management	
	SN 3.1.4	Coordinate joint and multinational training events		
	SN 8.1.1	Provide security assistance	Security-assistance management	
	SN 8.2.2	Support other government agencies		
	OP 4.7.1	Provide security assistance in the JOA	Security-assistance management	
	ART 8.3.3	Conduct security assistance	Security-assistance management	
Understand partner capabilities	SN 8.3	Coordinate military activities within the IA process		

Table E.1—Continued

Category	Task Number[a]	Task	6SOS	MiTT Training
Understand partner capabilities (continued)	ST 8.2.10	Coordinate multinational operations within theater	Political and cultural integration techniques	Cultural awareness
	ST 8.5.3.4	Coordinate planning for IA activities		
	OP 4.7.5	Coordinate politico-military support	CMO Political and cultural integration techniques (partner only)	Cultural awareness (partner only)
	OP 5.7.4	Coordinate plans with non-DoD organizations		
	NTA 4.8.3	Provide IA coordination		
	MCTL 5.4.3	Coordinate and integrate joint, multinational, and IA support	Political and cultural integration techniques (partner only)	Cultural awareness (partner only)
	SN 8.3.1	Coordinate and control policy for the conduct of operations		

Table E.1—Continued

Category	Task Number[a]	Task	6SOS	MiTT Training
Understand partner capabilities (continued)	OP 5.4.5	Coordinate and integrate component, theater, and other support	Political and cultural integration techniques	Cultural awareness
	OP 5.7.2	Determine national and agency capabilities and limitations	Regional orientation (partner only)	Cultural awareness (partner only)
	OP 4.7.6	Coordinate civil affairs in the JOA	CMO	
	SN 7.4.4	Conduct joint, multinational, interoperability, and IA training of assigned forces		
	ST 8.5.3.1	Establish JIACG to facilitate IA activities		
	NTA 4.8.4	Coordinate with NGOs		
	SN 7.4.6	Provide joint, multinational, interoperability, and IA training for other than assigned forces		

Table E.1—Continued

Category	Task Number[a]	Task	6SOS	MiTT Training
Understand partner capabilities (continued)	ST 8.4.2	Combat terrorism	Dynamics of international terrorism	
	ST 8.2.11	Coordinate with and support NGOs in theater		
	ST 8.5.3.3	Assess military participation during IA activities		
Conduct operations with and for partners	ART 8.3.2.3	Conduct combat operations in support of FID		
	NTA 6.3.3	Combat terrorism	Dynamics of international terrorism	
	ART 8.3.7.2	Conduct antiterrorism activities	Dynamics of international terrorism	
	MCTL 6.5.3	Combat terrorism	Dynamics of international terrorism	
	ART 8.3.5.1	Provide operational, logistic, and training support to insurgencies	Defense security–assistance management	
			Contemporary insurgent warfare	

Table E.1—Continued

Category	Task Number[a]	Task	6SOS	MiTT Training
Conduct operations with and for partners (continued)	OP 4.7.3	Provide support to DoD and other government agencies		
	SN 8.3.3	Establish IA cooperation structures		
	ART 6.14.6.18	Provide cultural-affairs support	Cross-cultural communication Regional orientation Political and cultural integration techniques	Cultural awareness
Collect and disseminate information	OP 2.2	Collect and share information		
	MCTL 2.1.1	Conduct area and country studies	Cross-cultural communication Regional orientation Political and cultural integration techniques	Cultural awareness
Support interpartner communications	MCTL 5.2.6	Provide liaison to local authorities and NGOs	CMO	

NOTE: Green = widely taught. Yellow = narrowly taught. Red = not taught.

[a] ST = strategic theater. OP = operational. SN = strategic national.

Highest Priorities for Integrated-Operations Training

Table F.1 describes those tasks that should take first priority in integrated-operations training.

Table F.1
Highest Priorities for Integrated-Operations Training

Task Number	Task	Task Description
ST 8.2.2	Coordinate civil affairs in theater	To coordinate those activities that foster relationships between theater military forces and civil authorities and people in a friendly country or area.
ART 7.10.3	Maintain community relations	To assist civil-affairs personnel in conducting (planning, preparing, executing, and assessing) community-relations programs as resources permit.
SN 3.1.3	Support establishment of access and storage agreements	To support the combatant commander's efforts to obtain agreements for periodic access by U.S. personnel and units and for the permanent stationing ashore or afloat of selected equipment and supplies.
OP 4.7.2	Conduct CMO in the JOA	To conduct activities that foster the relationship of the military forces with civilian authorities and population and that develop favorable emotions, attitudes, or behavior among neutral, friendly, or hostile groups.

Table F.1—Continued

Task Number	Task	Task Description
OP 4.4.5	Train joint forces and personnel	To train replacements and units, especially newly rebuilt units, in the theater of operations. In MOOTW, this activity includes training assistance for friendly nations and groups.
OP 4.7.7	Conduct FID	To provide assistance in the operational area to friendly nations facing threats to their internal security.
ART 6.14.6.2	Provide public-administration support	To provide liaison to the military forces. Survey and analyze the operation of local governmental agencies: their structure, centers of influence, and effectiveness.
ART 6.14.6.12	Provide public legal support	To establish supervision over local judiciary system, establish civil-administration courts, and help in preparing or enacting necessary laws for the enforcement of U.S. policy and international law.
ART 8.3.2.1	Provide indirect support to FID	Indirect support emphasizes the principles of HN self-sufficiency and builds strong national infrastructures through economic and military capabilities.
ART 8.3.2.2	Provide direct support to FID (not involving combat operations)	Direct support (not involving combat operations) involves the use of U.S. forces providing direct assistance to the HN civilian population or military.
ART 6.14.6.3	Provide public electronic-communication support	To manage communication resources, public and private, including postal services, telephone, telegraph, radio, television, and public-warning systems.
SN 8.1.2	Support-nation assistance	To support and assist in developing other nations, normally in conjunction with DOS or a multinational force (or both), and, ideally, through the use of HN resources.

Table F.1—Continued

Task Number	Task	Task Description
SN 8.1.6	Provide civil-affairs support policy	To provide policy on activities that embrace the relationship between a nation's military forces and its civil authorities and people in a friendly country or area or occupied country or area, when military forces are present.
ART 6.14.6.11	Provide civil-defense support	To ensure that an adequate civil-defense structure exists. Advise, assist, or supervise local civil-defense officials.
OP 4.7.1	Provide security assistance in the JOA	To provide friendly nations or groups with defense articles, military training, and other defense-related services by grant, loan, credit, or cash sales in furtherance of national policies and objectives within the JOA.
ART 3.3.2.4	Conduct PSYOP	To integrate planned psychological messages, products, and actions into combat operations or in support of MOOTW.
ST 8.5.1	Coordinate and integrate policy for the conduct of theater operations	To work within the country team and other forums to provide support to the programs of other USG departments and agencies within the theater.
MCTL 4.6.3	Plan, coordinate, and monitor EPWs, civilian internees, and U.S. military prisoner operations	To plan, coordinate, and monitor the collection, processing, and transfer of EPWs, civilian internees, and U.S. military prisoners.
ART 7.2.6	Communicate with non–English-speaking forces and agencies	To communicate orally, nonverbally, in writing, or electronically in the appropriate language of allied, HN, nongovernmental, and indigenous forces and agencies to accomplish all C2 requirements.
OP 2.1.2	Determine and prioritize operational IRs	To identify those items of information that must be collected and processed to develop the intelligence required by the commander's PIR.

Table F.1—Continued

Task Number	Task	Task Description
OP 5.1.5	Monitor the strategic situation	To be aware of and understand national and multinational objectives, policies, goals, other elements of national and multinational power, political aim, and the GCC's strategic concept and intent.
ART 7.10.1	Execute information strategies	To identify affected internal and external audiences and their information requirements.
NTA 6.1.2	Conduct perception management	To convey or deny selected information and indicators to foreign audiences to influence their emotions, motives, and objective reasoning.
ART 5.3.7.2	Conduct counterpropaganda	To establish plans and procedures to counter enemy PSYOP based on an effective PA and education program to expose, discount, and inform targeted audiences of threat-propaganda initiatives.
ART 7.4.2.2	Enhance friendly decisionmaking	To leverage information management that supports making more precise and timely decisions than the enemy.
NTA 6.3.1.6	Conduct surveillance-detection operations	To identify, locate, and help counter the enemy's intelligence, espionage, sabotage, subversion, and terrorism-related activities, capabilities, and intentions to deny the enemy the opportunity to take action against friendly forces.
ST 5.6	Develop and provide PA in theater	To develop and provide to the combatant commander and allied partners a program for telling the theater and combined command's story to audiences both internal and external.
ART 1.1.3	Provide intelligence support to force protection	To provide intelligence in support of protecting the tactical forces' fighting potential so that it can be applied at the appropriate time and place.

Table F.1—Continued

Task Number	Task	Task Description
ART 1.4.2.1.1	Provide intelligence support to PSYOP	This task identifies the cultural, social, economic, and political environment of the AO.
ART 1.4.2.3.1	Provide intelligence support to CMO	This task allows military intelligence organizations to collect and provide information and intelligence products concerning foreign cultural, social, economic, and political elements within an AO in support of CMO.
ART 3.3.2.4.1	Develop PSYOP products	To develop products to support offensive, defensive, stability, and support operations.
SN 8.1.7	Coordinate information-sharing arrangements	To arrange for the selected release and disclosure of unclassified and classified information in support of multinational operations and exercises.
MCTL 5.2.8	Coordinate PA services	To advise and assist the commander, associated commands, and coalition partners in providing information to internal and external audiences by originating and assisting civilian news media.
OP 5.5.1	Develop a joint-force C2 structure	To establish a structure for C2 of subordinate forces.

a ST = strategic theater. ART = article of the AUTL. SN = strategic national. NTA = article of the UNTL. MCTL = Marine Corps Task List.

Next-Highest Priorities for Integrated-Operations Training

Table G.1 describes the second-highest–priority tasks for training on integrated operations.

Table G.1
Next-Highest Priorities for Integrated-Operations Training

Task Number[a]	Task	Task Description
ST 8.3.1	Arrange stationing for U.S. forces	To obtain approval for and to house and dispose forces to best support peacetime presence and military operations.
ST 8.2.7	Assist in restoration of order	To halt violence and reinstitute peace and order.
OP 4.7.2	Conduct CMO in the JOA	To conduct activities that foster the relationship of the military forces with civilian authorities and population and that develop favorable emotions, attitudes, or behavior among neutral, friendly, or hostile groups.
OP 4.7.4	Transition to civil administration	To implement the transition from military administration in a region to UN or civil administration in the region.
ART 6.14.6	Establish temporary civil administration (friendly, allied, and occupied enemy territory)	To establish a temporary civil administration (at the direction of the NCA) until existing political, economic, and social conditions stabilize in enemy territory or in friendly territory where there is a weak or ineffective civil government.

Table G.1—Continued

Task Number[a]	Task	Task Description
SN 8.1.8	Provide support to FID in theater	To work with U.S. agencies and foreign governments to provide programs to support action programs to free and protect the foreign nation's society from subversion, lawlessness, and insurgency.
NTA 4.8	Conduct civil-affairs activities in area	To conduct those activities that embrace the relationship of the military forces with civil authorities and population in a friendly country or area or in an occupied country or area when military forces are present.
ART 6.13	Conduct internment and resettlement activities	Includes activities performed by units when they are responsible for interning EPWs and civilian detainees.
ST 8.2.1	Coordinate security-assistance activities	To provide defense articles, military training and advisory assistance, and other defense-related services.
SN 8.1.1	Provide security assistance	To provide defense articles, military training, and other defense-related services by grant, credit, or cash sales to further national policies and objectives.
SN 8.2.2	Support other government agencies	To support non-DoD agencies. Support includes military support to civil authorities and civilian LEAs, counterdrug operations, combating terrorism, noncombatant evacuation, and building a science and technology base.
ART 8.3.3	Conduct security assistance	Security assistance refers to a group of programs that support U.S. national policies and objectives by providing defense articles, military training, and other defense-related services to foreign nations by grant, loan, credit, or cash sales.

Table G.1—Continued

Task Number[a]	Task	Task Description
OP 4.7.6	Coordinate civil affairs in the JOA	To coordinate those activities that foster relationships of operational forces with local civil authorities and people in a friendly country or area.
ST 8.4.2	Combat terrorism	To produce effective anticipatory and offensive measures to defeat transnational terrorist organizations; prevent WMD acquisition, development, or use by terrorist organizations; and develop partner countries' capacity to detect and defeat terrorists.
ART 8.3.1.3	Conduct operations in support of diplomatic efforts	U.S. Army forces support diplomatic efforts to establish peace and order before, during, and after conflicts.
ART 6.13.2	Conduct populace and resource control	To provide security for a populace, denying personnel and materiel to the enemy, mobilize population and materiel resources, and detect and reduce the effectiveness of enemy agents.
ART 6.14.5	Resettle refugees and displaced civilians	To estimate the number of dislocated civilians, their points of origin, and anticipated direction of movement.
ART 7.7.2.2.5	Provide refugee and displaced-civilian movement control	To assist, direct, or deny the movement of civilians whose location, direction of movement, or actions may hinder operations.
OP 6.2	Provide protection for operational forces, means, and noncombatants	To safeguard friendly centers of gravity and operational-force potential by reducing or avoiding the effects of enemy operational level (tactical risks) actions.
SN 1.2.2	Provide forces and mobility assets	To provide the transportation assets (e.g., road, rail, sealift, and airlift) required in an operational configuration for the movement of forces and cargo.

Table G.1—Continued

Task Number[a]	Task	Task Description
MCTL 6.4.2	Conduct military law enforcement	To enforce military law and order and collect and evacuate EPWs.
ART 5.3.6.1	Provide protective services for selected individuals	To protect designated high-risk individuals from assassination, kidnapping, injury, or embarrassment.
ART 7.7.2.2	Provide law and order	To ensure lawful and orderly environment and suppress criminal behavior.
OP 4.6.4	Provide law enforcement and prisoner control	To collect, process, evacuate, and intern EPWs and to enforce military law and order in the COMMZ and in support of operational-level commander's campaigns and major operations.
ART 5.3.5.5	Conduct local security operations	To take measures to protect friendly forces from attack, surprise, observation, detection, interference, espionage, terrorism, and sabotage.
ART 7.7.2.2.1	Perform law enforcement	To assist commanders through conducting law-enforcement operations, including the maintenance of liaison activities and support of the training of other DoD police organizations and HN police authorities.
NTA 6.3.2	Conduct military law-enforcement support (afloat and ashore)	To enforce military law and order and collect, evacuate, and intern EPWs.
ART 8.3.2.2	Provide direct support to FID (not involving combat operations)	Direct support (not involving combat operations) involves the use of U.S. forces providing direct assistance to the HN civilian populace or military.
OP 2.1.1	Determine and prioritize operational PIRs	To assist USJFCOM in determining and prioritizing its PIRs. In MOOTW, it includes helping and training HNs to determine their IRs, such as in COIN operations.

Table G.1—Continued

Task Number[a]	Task	Task Description
OP 3.2.2.1	Employ PSYOP in the JOA	To plan and execute operations to convey selected information and indicators to foreign audiences in the AO to influence their emotions, motives, objective reasoning.
ART 6.14.6.18	Provide cultural-affairs support	To provide information to military forces on the social, cultural, religious, and ethnic characteristics of the local populace.
ART 7.10	Conduct PA operations	To advise and assist the commander and command (or HN, in MOOTW) in PA planning.
NTA 5.8	Provide PA services	To advise and assist the commander, associated commands, and coalition partners in providing information to internal and external audiences by originating print and broadcast news material and assisting with community-relations projects.
OP 6.2.12	Provide counter-PSYOP	To conduct activities to identify adversary psychological-warfare operations contributing to situational awareness and serve to expose adversary attempts to influence friendly populations and military forces.
MCTL 2.1.8	Provide tactical CI and HUMINT support	To sufficiently suppress or defeat the enemy's intelligence-collection, terrorism, and sabotage efforts to allow the regiment to conduct its mission with the element of surprise and with minimal losses.
MCTL 2.2	Collect information	To gather combat data and intelligence data to satisfy the identified requirements.
OP 5.1.3	Determine commander's critical information requirements	To determine the critical information that a commander requires to understand the flow of operations and to make timely and informed decisions.

Table G.1—Continued

Task Number[a]	Task	Task Description
ART 1.1.2	Perform situation development	Situation development is a process for analyzing information and producing current intelligence about the enemy and environment during operations.
ART 1.2.3	Conduct area studies of foreign countries	Study and understand the cultural, social, political, religious, and moral beliefs and attitudes of allied, HN, or indigenous forces to assist in accomplishing goals and objectives.
ART 1.4.2.3.2	Provide intelligence support to PA	This task identifies the coalition and foreign public physical and social environment, as well as world, HN national, and HN local public opinion.
SN 8.1.7	Coordinate information-sharing arrangements	To arrange for the selected release and disclosure of unclassified and classified information in support of multinational operations and exercises.
SN 8.1.3	Support peace operations	To support peace operations through national-level coordination of the three general areas: diplomatic action, traditional peacekeeping, and forceful military actions.
ART 6.14.1	Provide interface and liaison between U.S. military forces and local authorities and NGOs	To facilitate CMO by providing interface between U.S. military forces and HN, foreign authorities, foreign military forces, or NGOs.
MCTL 5.2.6	Provide liaison to local authorities and NGOs	To facilitate CMO by providing interface between U.S. military forces and HN, foreign authorities, or NGOs.

[a] ST = strategic theater. OP = operational. ART = article of the AUTL. SN = strategic national. NTA = article of the UNTL.

Tasks That Should Receive Less Training Emphasis

Table H.1 describes tasks that should receive less training emphasis than those first- and second-priority tasks described in the preceding two appendixes.

Table H.1
Tasks That Should Receive Less Training Emphasis

Task Number	Task	Task Description
SN 3.1.2	Coordinate periodic and rotational deployments, port visits, and military contacts	To collaborate with other U.S. departments and agencies and the U.S. Congress and to work with foreign governments to allow for U.S. combat, support, and training units; individual service members; and DoD civilians to visit foreign nations.
ST 8.1.1	Enhance regional politico-military relations	To strengthen and promote alliances through support of regional relationships.
SN 9.3	Conduct arms-control–support activities	To implement intrusive arms-control inspections to fulfill treaty obligations, including conducting on-site inspections.
SN 9.4	Support WMD nonproliferation and counterproliferation activities and programs	To implement and coordinate WMD nonproliferation and counterproliferation activities that respond to U.S. policy and strategy objectives for combating WMD proliferation and WMD terrorism.

Table H.1—Continued

Task Number	Task	Task Description
ST 7.1.7	Establish JMETL	To analyze applicable tasks derived through mission analysis of joint operation plans and external directives and select for training only those tasks that are essential to accomplish the organization's wartime mission.
ART 7.7.2.2.4	Provide customs support	To perform tactical actions that enforce restrictions on controlled substances and other contraband violations that enter or exit an AO.
ART 8.3.6.8	Provide research, development, and acquisition support to counterdrug efforts	The U.S. Army Counterdrug Research, Development, and Acquisition Office makes military research, development, and acquisition support available to LEAs.
ART 8.4.3.2.1	Provide support to domestic preparedness (for WMD)	The National Domestic Preparedness Office, under FEMA, orchestrates the national domestic-preparedness effort.
ART 8.4.3.3.3	Provide general support to civil law enforcement	Provide limited military support to LEAs. DoD may direct U.S. Army forces to provide training to federal, state, and local civilian LEAs.
ST 9.1	Integrate efforts to counter weapon and technology proliferation in theater	To integrate support of DoD and other government agencies to prevent, limit, or minimize the introduction of CBRNE weapons, new advanced weapons, and advanced weapon technologies to a region.
ST 9.2	Coordinate counterforce operations in theater	To positively identify and select CBRNE-weapon targets, such as acquisition, weaponization, facility preparation, production, infrastructure, exportation, deployment, and delivery systems.
SN 7.4.1	Coordinate JMETL or AMETL development	To provide methodology and policy for establishing combatant commander JMETL and combat-support AMETL.

Table H.1—Continued

Task Number	Task	Task Description
OP 5.5.3	Integrate joint-staff augmentees	To integrate augmentees into existing staff structure to form a joint staff to support USJFCOM.
OP 7.5	Integrate JOA ISR with CBRNE situation	To integrate the CBRNE-weapon situation into C4ISR systems in the JOA.
ART 5.3.7	Conduct defensive IO	To plan, coordinate, and integrate policies and procedures, operations, personnel, and technology to protect and defend information and information systems.
OP 5.5.4	Deploy joint-force HQ advance element	To deploy elements of the HQ into the operational area in advance of the remainder of the joint force.
OP 7.4	Coordinate CM in the JOA	To coordinate support for IA essential services and activities required to manage and mitigate damage resulting from the employment of CBRNE weapons or release of TIMs or contaminants.
OP 6.2.6	Conduct evacuation of noncombatants from the JOA	To use JOA military and HN resources for the evacuation of U.S. military dependents, USG civilian employees, and private citizens (U.S. and third-country nationals).
ART 2.1.1	Conduct mobilization of tactical units	Mobilization is the process by which U.S. Army tactical forces or part of them are brought to a state of readiness for war or other national emergency.
ART 2.1.1.1	Conduct alert and recall	Units and individuals receive mobilization or alert orders, individuals assigned to the unit are notified of the situation, and all individuals report to the designated location at the designated time with designated personal items.

Table H.1—Continued

Task Number	Task	Task Description
ART 2.1.1.2	Conduct home-station mobilization activities	This task involves the activities of reserve-component units at home station after receiving a mobilization order followed by entry onto federal active duty or other C2 changes.
ART 6.14.6.15	Provide food and agriculture support	To provide advice and assistance in establishing and managing crop-improvement programs, agricultural training, use of fertilizers and irrigation, livestock improvement, and food processing, storage, and marketing.
ART 8.3.8	Perform NEOs	NEOs relocate threatened civilian noncombatants from locations in a foreign nation to secure areas.
ART 8.3.10	Conduct a show of force	Shows of force are flexible deterrence options designed to demonstrate U.S. resolve.
NTA 1.4.8.1	Conduct alien-migrant interdiction operations	To intercept alien migrants at sea, rescue them from unsafe conditions, and prevent their passage to U.S. waters and territory.
NTA 4.4.2.4	Provide billeting to noncombatant evacuees	To use available military resources to provide accommodations, food, and emergency supplies to U.S. dependents, USG civilian employees, and private citizens (U.S. and third-nation) who have been evacuated from the AO.
MCTL 1.2.0.18	Conduct NEOs	Operations directed by DOS, DoD, or other appropriate authority, whereby noncombatants are evacuated from foreign countries when their lives are endangered by war, civil unrest, or natural disaster to safe havens or to the United States.

Table H.1—Continued

Task Number	Task	Task Description
SN 8.2.3	Support evacuation of noncombatants from theaters	To provide for the use of military and civil, including HN, resources for the evacuation of U.S. dependents, USG civilian employees, and private citizens (U.S. and third-nation).
SN 8.2.4	Assist civil defense	To assist other federal agencies and state governments in mobilizing, organizing, and directing the civil population to minimize the effects of enemy action or natural and technological disasters on all aspects of civil life.
SN 9.2.2	Coordinate CM	To contain, mitigate, and repair damage resulting from the intentional use or accidental release of a CBRNE weapon or a TIM.
SN 9.2.4.1	Support CM	To provide subject-matter expertise on site or through technical reachback to support CM, accident and incident response, and mitigation for domestic and foreign CM incidents and events.
ST 8.4.3	Coordinate evacuation and repatriation of noncombatants from theater	To use all available means to evacuate U.S. dependents, USG civilian employees, and private citizens (U.S. and third-country) from the theater and support the repatriation of appropriate personnel to the United States.
ST 8.4.5	Coordinate civil support in the United States	To plan for and respond to domestic requests for assistance from other USG and state agencies in the event of civil emergencies, such as natural and human-created disasters, CM, civil disturbances, and federal work stoppages.
NTA 6.2.1	Evacuate noncombatants from area	To use available military and civilian resources (including HN resources) to evacuate U.S. dependents, USG civilian employees, and private citizens (U.S. and third-nation) from the AO.

Table H.1—Continued

Task Number	Task	Task Description
ART 8.4.3.2.3	Respond to WMD incidents	Other government agencies have primary responsibility for responding to domestic terrorist and WMD incidents.
ART 8.4.3.3.1	Support U.S. Department of Justice counterterrorism activities	When directed by the NCA, U.S. Army forces may provide assistance to the Department of Justice in the areas of transportation, equipment, training, and personnel.
ART 8.4.3.3.2	Conduct civil-disturbance operations	The U.S. Army assists civil authorities in restoring law and order when state and local LEAs are unable to control civil disturbances.
ART 8.4.3.4	Provide community assistance	Community assistance is a broad range of activities that provide support and maintain a strong connection between the military and civilian communities.
TA 5.6	Employ TIO	TIO employed by joint services produce tactical information and gain, exploit, defend, or attack information or information systems.
ART 8.3.9	Conduct arms-control operations	Army forces normally conduct arms-control operations to support arms-control treaties and enforcement agencies.
OP 6.2.11	Provide counterdeception operations	To neutralize, diminish the effects of, or gain advantage from, a foreign deception operation.
OP 5.5.6	Establish or participate in task forces	To establish or participate in a functional or single-service task force established to achieve a specific limited objective. This task force may be single-service, joint, or multinational.

a SN = strategic national. ST = strategic theater. ART = article of the AUTL. NTA = article of the UNTL. TA = tactical.

Bibliography

1st Infantry Division officials, telephone interview with the authors, Fort Riley, Kan., January 2007.

1st Infantry Division, Military Transition Team Operation Iraqi Freedom–Operation and Enduring Freedom Training Model, Fort Riley, Kan., February 14, 2007.

6 Special Operations Squadron officials, interview with the authors, Hurlburt Field, Fla., December 2006.

AFSOC—*see* Air Force Special Operations Command.

Air Education and Training Command officials, interview with the authors, March 2007.

Air Force Security Assistance and Training Squadron officials, interview with the authors, San Antonio, Tex., March 2007.

Air Force Special Operations Command, Combat Aviation Advisor Mission Qualification Course Formal Training Pipeline, undated.

Battle Command Staff Training Program officials, interview with the authors, Fort Leavenworth, Kan., November 2006.

Brown, Bernice B., *Delphi Process: A Methodology Used for the Elicitation of Opinions of Experts*, Santa Monica, Calif.: RAND Corporation, P-3925, 1968. As of April 16, 2008:
http://www.rand.org/pubs/papers/P3925/

Bush, George W., Management of Interagency Efforts Concerning Reconstruction and Stabilization, national-security presidential directive 44, Washington, D.C., December 7, 2005.

Clinton, William J., Managing Complex Contingency Operations, presidential decision directive 56, Washington, D.C., May 1997.

DoD—*see* U.S. Department of Defense.

DOS—*see* U.S. Department of State.

Downie, Richard D., "Defining Integrated Operations," *Joint Force Quarterly*, Vol. 38, 3rd Quarter 2005, pp. 10–13. As of April 15, 2008: http://www.dtic.mil/doctrine/jel/jfq%5Fpubs/0538.pdf

FAS—*see* Federation of American Scientists.

Federation of American Scientists, "PDD/NSC 56: Managing Complex Contingency Operations," white paper, May 1997. As of April 15, 2008: http://www.fas.org/irp/offdocs/pdd56.htm

Glenn, Russell W., Jody Jacobs, Brian Nichiporuk, Christopher Paul, Barbara Raymond, Randall Steeb, and Harry J. Thie, *Preparing for the Proven Inevitable: An Urban Operations Training Strategy for America's Joint Force*, Santa Monica, Calif.: RAND Corporation, MG-439-OSD/JFCOM, 2006. As of April 17, 2008: http://www.rand.org/pubs/monographs/MG439/

Headquarters, U.S. Department of the Army, *The Army Universal Task List*, Washington, D.C., field manual 7-15, August 2003.

Interagency Transformation, Education, and Analysis Program, Interagency Coordination Symposium agenda, December 12–14, 2006.

Interagency Transformation, Education, and Analysis Program officials, interview with the authors, Fort McNair, Washington, D.C.

ITEA—*see* Interagency Transformation, Education, and Analysis Program.

JFSC—*see* Joint Forces Staff College.

Joint Center for International Security Force Assistance officials, interview with the authors, Fort Leavenworth, Kan., November 2006.

Joint Center for Operational Analysis officials, interview with the authors, Suffolk, Va., November 2006.

Joint Forces Staff College, Joint Interagency Multinational Planner's Course, JIMPC 07-1, November 13–17, 2006.

Joint Special Operations University, Special Operations Forces–Interagency Collaboration course, planning draft, July 21, 2006a.

———, Terrorism Response Senior Seminar Plan of Instruction, December 5–7, 2006b.

Joint War Fighting Center officials, interview with the authors, Washington, D.C., November 2006.

JSOU—*see* Joint Special Operations University.

Locher, James R., *Victory on the Potomac: The Goldwater-Nichols Act Unifies the Pentagon*, College Station, Tex.: Texas A&M University Press, 2002.

Marine Air-Ground Task Force, Staff Training Program officials, interview with the authors, Quantico, Va.

Marine Corps Training and Advisory Group, U.S. Marine Corps Forces Command briefing, October 2007.

Marine Corps University officials, interview with the authors, Quantico, Va.

MCTAG—*see* Marine Corps Training and Advisory Group.

Melillo, Michael R., "Outfitting a Big-War Military with Small-War Capabilities," *Parameters*, Vol. 36, No. 3, Autumn 2006, pp. 22–35. As of April 18, 2008: http://www.carlisle.army.mil/usawc/parameters/06autumn/melillo.pdf

Murdock, Clark A., *Beyond Goldwater-Nichols: Defense Reform for a Strategic Era: Phase One Report*, Washington, D.C.: Center for Strategic and International Studies, 2004.

Myers, Richard B., "A Word from the Chairman," *Joint Force Quarterly*, Vol. 36, 1st Quarter 2005, pp. 1–10. As of April 15, 2008: http://www.dtic.mil/doctrine/jel/jfq%5Fpubs/chmwd36.pdf

Nagl, LTC John, U.S. Army, discussion with the authors, undated.

National Defense Panel, and U.S. Department of Defense, *National Defense Panel Assessment of the May 1997 Quadrennial Defense Review*, Arlington, Va.: National Defense Panel, 1997.

National Defense University, *Interagency Management of Complex Crisis Operations Handbook*, January 2003. As of April 15, 2008: http://www.au.af.mil/au/awc/awcgate/ndu/interagency_mgt_crisis_ops_2003.pdf

Navy Warfare Development Command, *Universal Naval Task List*, version 3.0, Office of the Chief of Naval Operations instruction 3500.38B, January 2007.

———, *Navy Tactical Task List*, March 20, 2008.

NDU—*see* National Defense University.

NWDC—*see* Navy Warfare Development Command.

Office of the Under Secretary of Defense for Personnel and Readiness, "Department of Defense Training Transformation Implementation Plan FY2006–FY2011," February 23, 2006.

Phelan, Jake, and Graham Wood, *Bleeding Boundaries: Civil-Military Relations and the Cartography of Neutrality*, Surrey, UK: Ockenden International, November 2005. As of April 18, 2008: http://www.ockenden.org.uk/publications/pdf/BleedingBoundaries.pdf

Public Law 99-433, Goldwater-Nichols Department of Defense Reorganization Act, October 1, 1986.

Public Law 109-364, John Warner National Defense Authorization Act for Fiscal Year 2007, October 17, 2006.

Quinones, Miguel A., J. Kevin Ford, and Mark S. Teachout, "The Relationship Between Work Experience and Job Performance: A Conceptual and Meta-Analytic Review," *Personnel Psychology*, Vol. 48, Vol. 4, December 1995, pp. 887–910.

Security Cooperation Education and Training Center officials, interview with the authors, Arlington, Va., March 2007.

Special Operations Forces–Interagency Coordination Symposium, agenda, December 12–14, 2006.

Spiegel, Peter, "The World: Army Is Training Advisors for Iraq," *Los Angeles Times*, October 25, 2006, p. A1.

Sprenger, Sebastian, "Draft DoD Manual Could Bolster COCOM Leverage in Joint Training," *Inside the Pentagon*, January 18, 2007.

Terrorism Response Senior Seminar, plan of instruction, December 5–7, 2006.

Thie, Harry J., Margaret C. Harrell, and Robert M. Emmerichs, *Interagency and International Assignments and Officer Career Management*, Santa Monica, Calif.: RAND Corporation, MR-1116-OSD, 2000. As of April 17, 2008: http://www.rand.org/pubs/monograph_reports/MR1116/

UK Department for International Development officials, interview with the authors, London, UK, July 2006.

UK Ministry of Defense, Stabilisation Unit and Department of International Cooperation officials, interview with the authors, London, UK, July 2006.

U.S. Agency for International Development officials, interview with the authors, Washington, D.C., Fort Schafter, Hawaii, October 2006.

U.S. Army and U.S. Marine Corps Counterinsurgency Center officials, interview with the authors, Fort Leavenworth, Kan., November 2006.

U.S. Code, Title 10, Armed Forces. As of April 15, 2008: http://www.access.gpo.gov/uscode/title10/title10.html

———, Title 10, Section 3013, Secretary of the Army. As of April 16, 2008: http://frwebgate.access.gpo.gov/cgi-bin/getdoc. cgi?dbname=browse_usc&docid=Cite:+10USC3013

———, Title 10, Section 5013, Secretary of the Navy. As of April 16, 2008: http://frwebgate.access.gpo.gov/cgi-bin/getdoc. cgi?dbname=browse_usc&docid=Cite:+10USC5013

———, Title 10, Section 8013, Secretary of the Air Force. As of April 16, 2008: http://frwebgate.access.gpo.gov/cgi-bin/getdoc. cgi?dbname=browse_usc&docid=Cite:+10USC8013

U.S. Department of Defense, Department of Defense Participation in Integrated Operations, draft directive 3000.dd, undated.

————, Office of Public Affairs, *International Contributions to the War on Terrorism*, revised June 14, 2002a. As of April 16, 2008:
http://www.defenselink.mil/news/Jun2002/d20020607contributions.pdf

————, Unified Command Plan, revised October 1, 2002b.

———— Military Department Foreign Area Officer (FAO) Programs, directive 1315.17, April 28, 2005. As of April 17, 2008:
http://www.dtic.mil/whs/directives/corres/pdf/131517p.pdf

U.S. Department of State, *Training Continuum for Civil Service Employees*, 4th ed., undated(a). As of April 16, 2008:
http://fsitraining.state.gov/trainingCS.html

————, *Training Continuum for Foreign Service Generalists*, 2nd ed., undated(b). As of April 16, 2008:
http://fsitraining.state.gov/training/Training%20Continuum%20for%20
Foreign%20Service%20Generalists%20-%202nd%20Edition.pdf

————, *Leadership and Management Training Curriculum*, 2004.

————, Foreign Service Institute officials, interview with the authors, Arlington, Va., November 2006.

————, Office of the Coordinator for Reconstruction and Stability officials, interview with the authors, Washington, D.C., October 2006.

U.S. Government Accountability Office, *Military Training: Actions Needed to Enhance DoD's Program to Transform Joint Training; Report to Congressional Committees*, Washington, D.C., GAO-05-548, June 2005. As of April 18, 2008:
http://purl.access.gpo.gov/GPO/LPS61668

U.S. Joint Chiefs of Staff, *Joint Training Manual for the Armed Forces of the United States*, chair of the Joint Chiefs of Staff manual 3500.03A, September 1, 2002. As of April 16, 2008:
http://www.dtic.mil/doctrine/jel/cjcsd/cjcsm/m350003a.pdf

————, *Universal Joint Task List (UJTL)*, Washington, D.C., chair of the Joint Chiefs of Staff manual 3500.04D, August 1, 2005a.

————, *Officer Professional Military Education Policy (OPMEP)*, chair of the Joint Chiefs of Staff instruction 1800.01C, December 22, 2005b. As of April 16, 2008:
http://www.js.pentagon.mil/doctrine/education/cjcsi1800_01c.pdf

————, Joint Doctrine Division, *Department of Defense Dictionary of Military Terms*, joint publication 1-02, Washington, D.C., April 12, 2001, as amended through March 4, 2008. As of April 15, 2008:
http://purl.access.gpo.gov/GPO/LPS14106

————, Operational Plans and Joint Force Development (J7), Training and Exercises Division officials, interview with the authors, Pentagon, Arlington, Va., June 2006.

U.S. Joint Forces Command, "Multinational Experiment 5," undated Web page. As of April 16, 2008:
http://www.jfcom.mil/about/experiments/mne5.html

———, Strategy and Policy Directorate (J5), Joint Training Directorate and Joint Warfighting Center (J7/JWC), and Joint Concept Development and Experimentation Directorate officials, interview with the authors, Hampton, Va., April 2007.

U.S. Marine Corps, Center for Advanced Operational Culture Learning officials, interview with the authors, Quantico, Va., July 2006.

———, Combat Development Command, *Marine Corps Task List (MCTL)*, 2008. As of April 15, 2008:
https://www.mccdc.usmc.mil/MCTL.htm

U.S. Pacific Command, director for Strategic Planning and Policy (J5), Joint Interagency Coordination Group officials, interview with the authors, Honolulu, Hawaii.

USEUCOM—*see* U.S. European Command.

USJCS—*see* U.S. Joint Chiefs of Staff.

USJFCOM—*see* U.S. Joint Forces Command.

USMC—*see* U.S. Marine Corps.

Wong, Carolyn, *How Will the e-Explosion Affect How We Do Research? Phase I: The E-DEL+I Proof-of-Concept Exercise*, Santa Monica, Calif.: RAND Corporation, DB-399-RC, 2003. As of April 16, 2008:
http://www.rand.org/pubs/documented_briefings/DB399/